T0255244

Lecture Notes in Computer S

Edited by G. Goos, J. Hartmanis, and J. van Leeuwen

Springer

Berlin
Heidelberg
New York
Hong Kong
London
Milan
Paris
Tokyo

Ion Stoica

Stateless Core:
A Scalable Approach
for Quality of Service
in the Internet

Winning Thesis of
the 2001 ACM Doctoral Dissertation Competition

Springer

Series Editors

Gerhard Goos, Karlsruhe University, Germany
Juris Hartmanis, Cornell University, NY, USA
Jan van Leeuwen, Utrecht University, The Netherlands

Author

Ion Stoica
University of California at Berkeley, Computer Science Division
645 Soda Hall, Berkeley, CA 94720-1776, USA
E-mail: istoica@cs.berkeley.edu

Library of Congress Control Number: 2004104504

CR Subject Classification (1998): C.2, E.1, H.3.4-5, H.4.3

ISSN 0302-9743
ISBN 3-540-21960-9 Springer-Verlag Berlin Heidelberg New York

Springer-Verlag is a part of Springer Science+Business Media

springeronline.com

© Springer-Verlag Berlin Heidelberg 2004
Printed in Germany

Typesetting: Camera-ready by author, data conversion by Boller Mediendesign
Printed on acid-free paper   SPIN: 11007180   06/3142   5 4 3 2 1 0

To my wife Emilia and my son George

Foreword

The fundamental aspect of the Internet architecture that distinguishes it from other network technologies (such as X.25 and ATM) is that it is connectionless (vs. connection-oriented) and stateless (vs. stateful). The heated debate of whether connection-oriented or connectionless architecture is better has lasted for several decades. Proponents of the connectionless architecture point out the great robustness and scalability properties of the architecture, as demonstrated by the Internet. One well-known articulation of this philosophy is the "End-to-End Arguments". Opponents argue, rightfully, that there is no known solution that can provide quantitative performance assurances or guaranteed QoS in a connectionless network. It has been widely recognized that QoS is a must-have feature as the Internet technology evolves to the next stage. However, all existing solutions that provide guaranteed QoS require routers to maintain per flow (another name for connection used by the Internet community) state, which is the fundamental element of a connection-oriented architecture. The apparent conflicting goals of having a stateless network and supporting QoS have presented a great dilemma for Internet architects. As an example, Dave Clark, one of the most respected Internet architects and the author of the famous "End-to-End Arguments" paper, was also a key designer of the Internet Integrated Services Architecture that requires routers to maintain per flow state.

Dr. Ion Stoica's dissertation addresses this most pressing and difficult problem facing the Internet community today: how to enhance the Internet to support rich functionalities (such as QoS and traffic management) while still maintaining the scalability and robustness properties embodied in the original Internet architecture.

In his dissertation, Dr. Stoica proposes a novel architecture called SCORE (Stateless Core) that does not require core routers to maintain per flow state yet can provide services similar to those provided by stateful networks. This is achieved by a family of SCORE distributed algorithms that approximate the services provided by idealized stateful networks. The key technique used to implement a SCORE network is Dynamic Packet State (DPS), which uses extra state carried in packet headers to coordinate distributed algorithms implemented by routers. Such an architecture has both important theoretical and practical significances. From a conceptual point of view, this architecture is the first that combines the advantages of stateful and stateless networks, and can therefore achieve QoS, scalability, and robustness simultaneously. From a practical point of view, the industry and the IETF have been struggling to make a choice between two QoS architectures: the stateful Intserv, which can provide hard QoS guarantees but is less scalable and robust, and the stateless Diffserv, which is more scalable and robust but cannot provide services with high assurance. The SCORE architecture provides a third approach that is superior.

I believe that this research represents one of the most important and innovative contributions in networking research in the past decade. I hope that you will enjoy reading it and agree with me afterward.

Hui Zhang

Preface

This book contains the dissertation the author wrote at the Department of Electrical and Computer Engineering (ECE) at Carnegie Mellon University. This thesis was submitted to the ECE department in conformity with the requirements for the degree of Doctor of Philosophy in 2000. It was honored with the 2001 ACM Doctoral Dissertation Award.

Abstract

Today's Internet provides one simple service: best-effort datagram delivery. This minimalist service allows the Internet to be *stateless*, that is, routers do not need to maintain any fine-grained information about traffic. As a result of this stateless architecture, the Internet is both highly *scalable* and *robust*. However, as the Internet evolves into a global commercial infrastructure that is expected to support a plethora of new applications, such as IP telephony, interactive TV, and e-commerce, the existing best-effort service will no longer be sufficient. As a consequence, there is an urgent need to provide more powerful services such as guaranteed services, differentiated services, and flow protection.

Over the past decade, there has been intense research toward achieving this goal. Two classes of solutions have been proposed: those maintaining the *stateless* property of the original Internet (e.g., differentiated services), and those requiring a new *stateful* architecture (e.g., integrated services). While stateful solutions can provide more powerful and flexible services, such as per flow bandwidth and delay guarantees, they are less scalable than stateless solutions. In particular, stateful solutions require each router to maintain and manage per flow state on the control path, and to perform per flow classification, scheduling, and buffer management on the data path. Since today's routers can handle millions of active flows, it is difficult, if not impossible, to implement such solutions in a scalable fashion. On the other hand, while stateless solutions are much more scalable, they offer weaker services.

The key contribution of this dissertation is to bridge this long-standing gap between stateless and stateful solutions in packet-switched networks such as the Internet. Our thesis is that *"it is actually possible to provide services as powerful and as flexible as the ones implemented by a stateful network using a stateless network*". To prove this thesis, we propose a novel technique called Dynamic Packet State (DPS). The key idea behind DPS is that, instead of having routers maintain per flow state, packets carry the state. In this way, routers are still able to process packets on a per flow basis, despite the fact that they do not maintain any per flow state. Based on DPS, we develop a network architecture called Stateless Core (SCORE) in which core routers do not maintain any per flow state. Yet, using DPS to coordinate actions of edge and core routers along the path traversed by a flow allows us to design distributed algorithms that emulate the behavior of a broad class of stateful networks in SCORE networks.

In this dissertation we describe complete solutions including architectures, algorithms, and implementations which address three of the most important problems in today's Internet: providing guaranteed services, differentiated services, and flow protection. Compared to existing solutions, our solutions *eliminate* the most complex operations on both the data and control paths in the network core, i.e., packet classification on the data path, and maintaining per flow state consistency on the control path. In addition, the complexities of buffer management and packet scheduling are greatly reduced. For example, in our flow protection solution these operations take constant time, while in previous solutions these operations may take time logarithmic in the number of flows traversing the router.

Acknowledgements

I am extremely grateful to Hui Zhang, my thesis advisor, for giving me the right amount of freedom and guidance during my graduate studies. From the very beginning, he treated me as a peer and as a friend. He was instrumental in maintaining my focus, and constantly reminding me that identifying the research problem is as important, if not more important, than finding the right solution. Hui not only taught me how to become a better researcher, but also helped me to become a better person. His engaging arguments and strong feedback contributed greatly to this dissertation. I hope for and look forward to continued collaboration with him in the future.

The genesis of this thesis can be traced back to my internship at Xerox PARC in the summer of 1997. It all started with Scott Shenker asking the intriguing question: "Can we approximate Fair Queueing without maintaining per flow state in a network cloud?" I am indebted to Scott for teaching me how to rigorously define a problem and then pursue its solution. During these years he was an invaluable source of feedback and support. His suggestions and insights had a great impact on this dissertation.

I am grateful to the other members of my committee, Garth Gibson, Thomas Gross, and Peter Steenkiste, for their feedback and for their advice that helped to shape my research skills. Garth taught me the art of asking the right questions in a technical discourse. His inquisitorial and sharp questions helped me to better understand the limitations of my research and motivated me to find better ways to explain my results. Thomas provided the right balance to my research by constantly encouraging me to not get buried in the algorithmic details, but to always try to put my work into perspective. Peter always found time to discuss research issues, and gave excellent feedback. He was one of the first to suggest using Dynamic Packet State to provide guaranteed services.

Thanks to Mihai Budiu, Yang-hua Chu, Urs Hengartner, Eugene Ng, Mahadev Satyanarayanan, and Jeannette Wing for their feedback and comments that helped to improve the overall quality of this dissertation. Thanks to Joan Digney for accommodating me in her busy schedule and proofreading this thesis, which helped to significantly improve the presentation.

I am grateful to all the friends and colleagues with whom I spent my time as a graduate student at Carnegie Mellon University. Thanks to my Romanian friends, Mihai and Raluca Budiu, Cristian Dima, Marius Minea, and George Necula, with whom I spent many hours discussing the most varied and exciting topics. Mihai sparkled many of these conversations with his wit, and by being a never empty reservoir of information. Thanks to my networking group colleagues and officemates Yang-hua Chu, Urs Hengartner, Nick Hopper, Tom Kang, Marco Mellia, Andrew Myers, Eugene Ng, Sanjay Rao, Chuck Rosenberg, Kay Sripanidkulchai, Donpaul Stephens, and Yinglian Xie. I treasure our animated Friday lunch discussions that always managed to end the week on a high note. Special thanks to Eugene for his help and patience with my countless questions. He was the default good-to-answer-all-questions person whom I asked about everything, from how to modify the `if_de` driver in FreeBSD, to which are the best restaurants around the campus.

The completion of this thesis marked the end of my many years as a student. Among many outstanding teachers who shaped my education and scientific career are: Hussein Abdel-Wahab, Irina Athanasiu, Kevin Jeffay, David Keyes, Trandafir Moisa, Stephan Olariu, Alex Pothen, and Nicolae Tapus. I am grateful to Kevin, who was instrumental in helping and then convincing me to come to Carnegie Mellon University. Many thanks to Irina who, during my studies at the "Politehnica" University of Bucharest, gave me the mentorship a student can only dream about.

I would like to express my earnest gratitude to my parents and my sister for their love and support, without which any of my achievements would not have been possible. Thanks to my father who sparkled my early interest in science and engineering. His undaunting confidence gave me the strength to overcome any difficulties and to maintain high goals. Thanks to my mother for her love and countless sacrifices to raise me and to give me the best possible education.

I am deeply indebted to my dear wife Emilia for her love and understanding through my graduate years. She was always behind me and gave her unconditional support even if that meant sacrificing the time we spent together. I thank my mother and my mother-in-law who devotedly took care of our son for two and a half years. Without their help, it would not have been possible to reach this stage in my career. Finally, thanks to our son, George, for the joy and the happiness he brings to me during our many moments together.

Berkeley, December 2003   *Ion Stoica*

Contents

1 Introduction

Today's Internet provides one simple service: best effort datagram delivery. Such a minimalist service allows routers to be *stateless*, that is, except for the routing state, which is highly aggregated, routers do not need to maintain any fine grained state about traffic. As a consequence, today's Internet is both highly *scalable* and *robust*. It is scalable because router complexity does not increase in either the number of flows or the number of nodes in the network, and it is robust because there is little state, if any, to update when a router fails or recovers. The scalability and robustness are two of the most important reasons behind the success of today's Internet.

However, as the Internet evolves into a global commercial infrastructure, there is a growing need to provide more powerful services than best effort such as guaranteed services, differentiated services, and flow protection. Guaranteed services would make it possible to guarantee performance parameters such as bandwidth and delay on a per flow basis. An example would be to guarantee that a flow receives at least a specified amount of bandwidth, ensuring that the delay experienced by its packets does not exceed a specified threshold. This service would provide support for new applications such as IP telephony, video-conferencing, and remote diagnostics. Differentiated services would allow us to provide bandwidth and loss rate differentiation for traffic aggregates over multiple granularities ranging from individual flows to the entire traffic of a large organization. An example would be to allocate to one organization twice as much bandwidth on every link in the network as another organization. Flow protection would allow diverse end-to-end congestion control schemes to seamlessly coexist in the Internet, protecting the well behaved traffic from the malicious or ill-behaved traffic. For example, if two flows share the same link, with flow protection, each flow will get at least half of the link capacity independent of the behavior of the other flow, as long as the flow has enough demand. In contrast, in today's Internet, a malicious flow that sends traffic at a higher rate than the link capacity can provoke packet losses to another flow no matter how little traffic that flow sends!

Providing these services in packet switched networks such as the Internet has been one of the major challenges in the network research over the past decade. To address this challenge, a plethora of techniques and mechanisms

I. Stoica: Stateless Core, LNCS 2979, pp. 1-11, 2004.

have been developed for packet scheduling, buffer management, and signaling. While the proposed solutions are able to provide very powerful network services, they come at a cost: *complexity*. In particular, these solutions usually assume a *stateful* network architecture, that is, a network in which every router maintains per flow state. Since there can be a large number of active flows in the Internet, and this number is expected to continue to increase at an exponential rate, it is an open question whether such an architecture can be efficiently implemented. In addition, due to the complex algorithms required to set and preserve the state consistency across the network, robustness is much harder to achieve.

In summary, while stateful architectures can provide more sophisticated services than the best effort service, stateless architectures such as the current Internet are more scalable and robust. The natural question is then: Can we achieve the best of the two worlds? That is, *is it possible to provide services as powerful and flexible as the ones implemented by a stateful network in a stateless network?*

In this dissertation we answer this question affirmatively by showing that some of the most representative Internet services that require stateful networks can indeed be implemented in a mostly stateless network architecture.

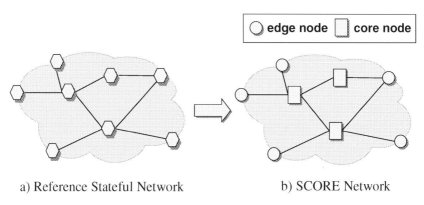

a) Reference Stateful Network   b) SCORE Network

Fig. 1.1. (a) A reference stateful network whose functionality is approximated by (b) a Stateless Core (SCORE) network. In SCORE only edge nodes maintain per flow state and perform per flow management; core nodes do not maintain any per flow state.

1.1 Main Contribution

The main contribution of this dissertation is to *provide the first solution that bridges the long-standing gap between stateless and stateful network architectures.* In particular, we show that three of the most important Internet services proposed in literature during the past decade, and for which the previous known solutions require stateful networks, can be implemented in a stateless core network. These services are: (1) guaranteed services, (2) service differentiation for large granularity traffic, and (3) flow protection to provide network support for congestion control.

The main goal of our solution is to push the state and therefore the complexity out of the network core, *without* compromising network ability to provide per flow services. The key ideas that allow us to achieve this goal are:

1. instead of having core nodes maintain per flow state, have packets carry this state, and
2. use the state carried by the packets to implement distributed algorithms to provide network services as powerful and as flexible as the ones implemented by stateful networks

The following paragraphs present the main components of our solution:

The Stateless Core (SCORE) Network Architecture The basic building block of our solution is the Stateless Core (SCORE) domain. We define a SCORE domain as being a trusted and contiguous region of network in which only edge routers maintain per flow state; the core routers do *not* maintain any per flow state (see Figure 1.1(b)). Since edge routers usually run at a much lower speed and handle far fewer flows than core routers, this architecture is highly scalable.

The "State-Elimination" Approach Our ultimate goal is to provide powerful and flexible network services in a stateless network architecture. To achieve this goal, we propose an approach, called "state-elimination" approach, that consists of two steps (see Figure 1.1). The first step is to define a reference stateful network that implements the desired service. The second step is to approximate or, if possible, to emulate the functionality of the reference network in a SCORE network. By doing this, we can provide services as powerful and flexible as the ones implemented by a stateful network in a mostly stateless network architecture, i.e., in a SCORE network.

The Dynamic Packet State (DPS) Technique To implement the approach, we propose a novel technique called Dynamic Packet State (DPS). As shown in Figure 1.2, with DPS, each packet carries in its header some state that is initialized by the ingress router. Core routers process each incoming packet based on the state carried in the packet's header, updating both its internal state and the state in the packet's header before forwarding it to the next hop. In this way, routers are able to process packets on a per flow basis, despite the fact that they do not maintain per flow state. As we will demonstrate in this dissertation, by using DPS to coordinate the actions of edge and core routers along the path traversed by a flow, it is possible to

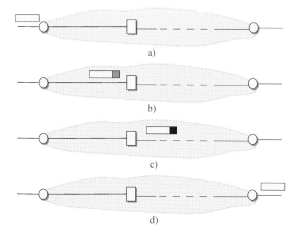

Fig. 1.2. An illustration of the Dynamic Packet State (DPS) technique used to implement per flow services in a SCORE network: (a-b) upon a packet arrival the ingress node inserts some flow dependent state into the packet header; (b-c) a core node processes the packet based on this state, and eventually updates both its internal state and the packet state before forwarding it. (c-d) the egress node removes the state from the packet header.

design distributed algorithms to approximate the behavior of a broad class of stateful networks using networks in which core routers do not maintain per flow state.

The "Verify-and-Protect" Approach   While our solutions based on SCORE/DPS have many advantages over traditional stateful solutions, they still suffer from robustness and scalability limitations when compared to stateless solutions. The scalability of the SCORE architecture suffers from the fact that the network core cannot transcend trust boundaries (such as boundaries between competing Internet Service Providers), and therefore high-speed routers on these boundaries must be stateful edge routers. System robustness is limited by the possibility that a single edge or core router may malfunction, inserting erroneous information in the packet headers, thus severely impacting performance of the entire network.

In Chapter 7 we propose an approach, called "verify-and-protect", that overcomes these limitations. We achieve scalability by pushing the complexity all the way to the end-hosts, eliminating the distinction between edge and core routers. To address the trust and robustness issues, all routers statistically verify that the incoming packets are correctly marked. This approach enables routers to discover and isolate misbehaving end-hosts and routers.

1.2 Other Contributions

To achieve the goal of providing the same level of services in a SCORE network as in traditional stateful networks, we propose several novel distributed algorithms that use DPS to coordinate the actions between the edge and core nodes. Among these algorithms are:

Core-Stateless Fair Queueing (CSFQ) This is the first algorithm to approximate the bandwidth allocation achieved by a stateful network in which all routers implement Fair Queueing [31, 79] in a core stateless network. As discussed in Chapter 4, CSFQ allows us to provide per flow protection in a SCORE network.

Core Jitter Virtual Clock (CJVC) This is the first algorithm to provide the same *worst-case* bandwidth and delay guarantees as Jitter Virtual Clock [126] and Weighted Fair Queueing [31, 79] in a network architecture in which core routers maintain no per flow state. CJVC implements the full functionality on the data path to provide guaranteed services in a SCORE network (see Chapter 5).

Distributed admission control We propose a distributed per flow admission control protocol in which core routers need to maintain *only* aggregate reservation state. To maintain this state, we develop a robust algorithm based on DPS that provides the same or even stronger semantics than those provided by previously proposed stateful solutions such as the ATM User-to-Network (UNI) signaling protocol and Reservation Protocol (RSVP) [1, 128]. Admission control is a key component of providing guaranteed services. It allows us to reserve bandwidth and buffer space at each router along a flow path to make sure that flow bandwidth and delay requirements are met.

Route pinning We propose a light-weight protocol and mechanisms to bind a flow to a specific route (path) through a network domain, without requiring core routers to maintain per flow state. This can be viewed as an alternative to Multi-Protocol Label Switching (MPLS) [17]. Our solutions for guaranteed and differentiated services use route pinning to make sure that all packets of a flow traverse the same path (see Chapters 5 and 6).

A major challenge in implementing the DPS-based algorithms is to find extra space in the packet header to encode the per flow state. Since this extra space is at premium, especially in the context of IPv4, we need to encode the state as efficiently as possible. To address this problem, we introduce two general methods to achieve efficient state encoding.

In the first method, the idea is to leverage knowledge about the state semantics. In particular, to save space we can use this knowledge to store a value as a function of another value. For example, if a value is known to be always greater than another value, we can use an accurate floating point representation to represent the larger value, and store the smaller value as a fraction of the larger one.

The idea behind the second method is to have different packets of a flow carry different state formats. This method is appropriate for algorithms that do not require all packets to carry the same type of state. For example, an algorithm may use the same field in the packet header to insert either data or control path information, as long as this will not compromise the service semantics.

1.3 Evaluation

In order to evaluate the solutions proposed in this dissertation, we try to answer the following three questions:

1. How scalable are the algorithms implemented by core routers? *Scalability* represents the ability of the network to grow in the number of flows (users), the number of nodes, and the traffic volume. To answer this question, we express the complexity of the proposed algorithms as a function of these parameters. In particular, we will show that our DPS based algorithms implemented by core routers are highly scalable as their complexity does not depend on either the number of flows or the network size.
2. How close is the service provided by our solution to the service provided by the reference stateful network? A service is usually defined in terms of performance parameters such as bandwidth, delay and loss rate. We answer this question by comparing the performance parameters achieved under our solution and the reference stateful solution. For example, in the case of the guaranteed services, we will show that end-to-end delay bounds of a flow in our core stateless solution are identical to the end-to-end delay bounds of the same flow in the reference stateful solution (see Section 5.3.3).
3. How does the service provided by our solution compare to similar services provided by existing stateless solutions? Again, we answer this question by comparing the performance parameters of services provided by our solution and the stateless solutions. However, unlike the previous question where the goal is to see how well we emulate the target service implemented by a reference stateful network, in this case, our goal is to see how much we gain in terms of service quality in comparison to existing stateless solutions. For example, in the case of flow protection, we will show that none of the traditional solutions that exhibit the same complexity at core routers is effective in providing flow protection (see Section 4.4).

To address the above three questions, we use a mix of theoretical analysis, simulations, and experimental results. In particular, to answer the first question, we use theoretical analysis to derive the time and space complexity of the algorithms performed by both edge and core routers. To answer the last two questions we derive worst-case or asymptotic bounds for the performance parameters that characterize the service, such as delay and bandwidth. Whenever we cannot obtain such bounds, or if we want to relax the assumptions to fit more realistic scenarios, we rely on extensive simulations by using an accurate packet level simulator such as ns-2 [78].

For illustration, consider our solution to provide per flow protection in a SCORE network (see Chapter 4). To answer the scalability question we show that in our solution a core router does not need to maintain any per flow state, and that the time it takes to process a packet is independent of the number of flows that traverse the router, n. In contrast, with the existing solutions, each router needs to maintain state for every flow, and the time it takes to process a packet increases with $\log n$. Consequently, our solution exhibits an $O(1)$ space and time complexity, as compared to existing solutions that exhibit an $O(n)$ space complexity, and an $O(\log n)$ time complexity. To answer the second question we use theoretical analysis to show that the difference between the average bandwidth allocated to a flow in a SCORE network and the bandwidth allocated to the same flow in the reference network is bounded. In addition, to answer the third question and to study more realistic scenarios, we use extensive simulations.

Finally, to demonstrate the viability of our solutions and to explore the compatibility of the DPS technique with IPv4, we present a detailed implementation in FreeBSD, as well as experimental results, to evaluate accuracy and implementation overhead.

1.4 Discussion

In this dissertation, we make two central assumptions. The first is that the ability to process packets on a per flow basis is beneficial, and perhaps even crucial, for supporting new emerging applications in the Internet. The second is that it is very hard, if not impossible, for traditional stateful solutions to support these services in high-speed backbone routers. It is important to note that these two assumptions do not necessary imply that it is infeasible to support these emerging services in high speed networks. They just illustrate the drawback of existing solutions that require routers to maintain and manage per flow state. In this dissertation we eliminate this problem, by demonstrating that it is possible to process packet on a per flow basis without requiring high-speed routers to maintain any per flow state.

The next two sections motivate these assumptions.

1.4.1 Why Per Flow Processing?

The ability to process packets on a per flow basis is important because it would allow us simultaneously (1) to support applications with different performance requirements, and (2) to achieve high resources utilization. To illustrate this point consider a simple example in which a file transfer application and an audio application share the same link. On one hand, we want the file transfer application to be able to use the entire link capacity, when the audio source does not send any traffic. On the other hand, when the audio application starts the transmission, we want this application to be able immediately to reclaim its share of the link capacity. In addition, since the audio application is much more sensitive to packet delay than the file transfer application, we should be able to preferentially treat the audio traffic in order to minimize its delay. As demonstrated by previous proposals, such a functionality can be easily provided in a stateful network in which routers process packets on a per flow basis [10, 48, 106].

A natural question to ask is whether performing packet processing at a coarser granularity, that is, on a per class basis, wouldn't allow us to achieve similar results. With such an approach, applications with similar performance requirements would be aggregated in the same traffic class. This would make routers much simpler to implement, as they need to differentiate between potentially only a small number of classes, rather than a large number of flows. While this approach can go a long way to support new applications in a scalable fashion, it has fundamental limitations. The main problem is that this approach implicitly assumes that *all* applications in the same class (1) cooperate, and (2) have similar requirements at *each* router. If assumption (1) does not hold, then malicious users may arbitrarily degrade the service of other users in the same class. If assumption (2) does not hold, it is very hard to meet all application requirements and simultaneously achieve efficient resource utilization. Unfortunately, these assumptions do not necessarily hold in practice. As we discuss in Chapter 4, cooperation is hard to achieve in today's Internet: even in the absence of malicious users, there is a natural incentive for a user to aggressively send more and more traffic in the hope of making other users quit and grabbing their resources. Assumption (2) may not hold simply because applications care about the end-to-end performance, and not about the local performance they experience at a particular router. As a result, applications with similar end-to-end performance requirements may end up having very different performance requirements at individual routers. For example, consider two flows that carry voice traffic and belong to the same class, one traversing a 15 node path, and another traversing a three node path. In addition, assume that, as suggested by recent studies in the area of interactive voice communication [7, 64], the tolerable end-to-end delay for both flows is about 100 ms, and that the propagation delay alone along each path is 10 ms. Then, while the first flow can afford a delay of only 6 ms per router, the second flow can afford a delay of up to 30 ms per router. But if both flows traverse the same router, the router will have to provide a 6 ms delay to both flows, as it does not have any way to differentiate between the two flows. Unfortunately, as we show in Appendix B.1, even under very low link utilization (e.g., 15%), it is very difficult to provide small delay bounds for all flows.

In summary, the ability to process packets on a per flow basis is highly desirable not only because it allows us to support applications with diverse needs, but also because it allows us to maximize the resource utilization by closely matching the application requirements to resource consumption.

1.4.2 Scalability Concerns with Stateful Network Architectures

In this section, we argue that the existing solutions that enable packet processing on a per flow basis, that is, stateful solutions, have serious scalability limitations, and that these limitations make the deployment of these solutions unlikely in the foreseeable future.

Recall that by scalability we mean the ability of a network to grow in the number of nodes, in the number of users it can support, and the traffic volume it can carry. Since in today's Internet these parameters increase at an exponential rate, scalability is a fundamental property of any protocol or algorithm to be deployed in the Internet. Indeed, according to recent statistics, Internet traffic doubles every six months, and it is expected to do so until 2008 [88]. This growth is fueled by both the exponential increase in the number of hosts, and the increase of bandwidth available to end users. The estimated number of hosts[1] reached 72 million in February 2000, and it is expected to reach 1 billion by 2008 [89]. In addition, the replacement of the ubiquitous 56 Kbps modems with cable modems and Digital Subscriber Line (DSL) connections will increase home users' access bandwidth by at least one order of magnitude.

In spite of such a rapid growth, a question still remains: with the continuous increase in available processor speed and memory capacity, wouldn't it be feasible to implement stateful solutions at very high speeds? In the remainder of this section, we answer this question. In particular, we first discuss why it is hard to implement per flow solutions today, and then we argue that it will be even harder to implement them in the foreseeable future.

Very high-end routers today can switch on the order of terabits per second, and handle individual links of up to 20 Gbps [2]. With an average packet size of 500 bytes, an input has only 25 ns to process a packet. If we assume a 1 GHz processor that is capable of executing an instruction every clock cycle, we have have just 25 instructions available per packet. During this time a router has to read the packet header, classify the packet to the flow it belongs to based on the fields in the packet header, and then process the packet based on the state associated to the flow. Packet processing may include rate regulation, and packet scheduling based on some arbitrary parameter such as the packet

---

[1]This number represents only hosts with Domain Names. The actual number of computers that are connected to the Internet is much larger, but this number is much more difficult to estimate.

deadline. In addition, stateful solutions requires the set up of per flow state, and the maintenance of this state consistency at all routers on the flow's path. Maintaining the state consistency in a distributed network environment such as the Internet in which packets can be lost or arbitrary delayed, and routers can fail is a very difficult problem [4, 117]. Primarily due to these technical difficulties, none of the high-end routers today implement stateful solutions.

While throwing more and more transistors at the problem will help, this will not necessarily solve the problem. Even if, as Moore's law predicts, processor performance continues to double every 18 month, this increase may not be able to offset the faster increase of the Internet traffic volume, which doubles every six moths. Worse yet, the increase in the router capacity not only reduces the time available to process a packet, but can also increase the amount of work the router has to do per packet. This is because a higher speed router will handle more flows, and the complexity of some of the per packet operations, such as packet classifications, and scheduling, depends on the number of flows. Even factoring out the algorithmic complexity, maintaining per flow state has the disadvantage of requiring a large memory footprint, which will negatively impact the memory access times. Finally, the advances in semiconductor performances will do little to address the challenge of maintaining the per flow state consistency, arguably the most difficult problem faced by today's proposals to provide per flow services.

1.5 Organization

The rest of this dissertation is organized as follows: Chapter 2 provides background information. In the first part, it presents the IP network model which is the foundation of today's Internet. In the second part, it discusses two of the most prominent proposals to provide better service in the Internet: Integrated Services and Differentiated Services. The chapter emphasizes the trade-offs between providing stronger semantics services and implementation complexity.

Chapter 3 describes the main components of our solution, and gives three simple examples to illustrate the DPS technique. The solution is then compared in terms of scalability and robustness against traditional solutions aiming to provide similar services in the Internet.

Chapters 4, 5, and 6 describe three important network services that can be implemented by our solution: (1) flow protection to provide network support for congestion control, (2) guaranteed services, and (3) service differentiation for large traffic aggregates, respectively. Our solution is the first to implement flow protection for congestion control and guaranteed services in a stateless core network architecture. We use simulations or experimental results to evaluate our solutions and compare them to existing solutions that provide similar services.

Chapter 7 describes a novel approach called "verify-and-protect" to overcome some of the scalability and robustness limitations of our solution. We illustrate this approach in the context of providing per flow protection, by developing tests to accurately identify misbehaving nodes, and present simulation results to demonstrate the effectiveness of the approach.

Chapter 8 presents our prototype implementation which provides guaranteed services and per flow protection. It discusses compatibility issues with the IPv4 protocol, and the information encoding in the packet header. The latter part of the chapter discusses a light weight monitoring tool that is able to continuously monitor the traffic on a per flow basis without affecting real-time guarantees.

Finally, Chapter 9 summarizes the conclusions of the dissertation, discusses the limitations of our work, and ends with directions for future work.

2 Background

Over the past decade, two classes of solutions have been proposed to provide better network services than the existing best effort service in the Internet: those maintaining the stateless property of the original Internet (e.g., Differentiated Services), and those requiring a new stateful architecture (e.g., Integrated Services). While stateful solutions can provide more powerful and flexible services such as per flow guaranteed services, and can achieve higher resource utilization, they are less scalable than stateless solutions. On the other hand, while stateless solutions are much more scalable, they offer weaker services. In this chapter, we first present all the mechanisms that a router needs to implement in order to support these solutions, and then discuss in detail the implementation complexity of each solution and the service quality it achieves.

The remainder of this chapter is organized as follows. Section 2.1 discusses the two main communication models proposed in the literature: circuit switching and packet switching. Section 2.2 presents the Internet Protocol (IP) network model, the foundation of today's Internet. In particular, the section discusses the operations performed by existing and the next generation routers on both the data and control paths. Data path consists of all operations performed by a router on a packet as the packet is forwarded to its destination, and includes packet forwarding, packet scheduling, and buffer management. Control path consists of the operations and protocols used to initialize and maintain the state required to implement the data path functionalities. Examples of control path operations are constructing and maintaining the routing tables, and performing admission control. Section 2.3 presents a taxonomy of services in a packet switching network. Based on this taxonomy, we discuss some of the most prominent services proposed in the context of the Internet: the best effort service, flow protection, Integrated Services, and Differentiated Services. We then compare these solutions in terms of the quality of service they provide and their complexity. Section 2.4 concludes this chapter by summarizing our findings.

I. Stoica: Stateless Core, LNCS 2979, pp. 13-33, 2004.
© Springer-Verlag Berlin Heidelberg 2004

2.1 Circuit Switching Vs. Packet Switching

Communication networks can be classified into two broad categories: packet switching and circuit switching. Circuit switching networks are best represented by telephone networks, first developed more than 100 years ago. In these networks, when two end points need to communicate, a dedicated channel (circuit) is set up between them. The channel remains open for the entire duration of the call, no matter whether the channel is actually used or not.

Packet switching networks are best exemplified by the Asynchronous Transport Mode (ATM) and Internet Protocol (IP) networks. In these networks information is carried by packets. Each packet is switched and transmitted through the network based on the information contained in the packet header. At the destination, the packets are reassembled to reconstruct the original information.

The most important advantage of packet switching over circuit switching is the ability to exploit *statistical multiplexing*. Unlike circuit switching where no one can use an open channel if its end-points do not use it, with packet switching, active sources can use any excess capacity made available by the inactive sources. In a networking environment with bursty traffic, allowing sources to share network resources can significantly increase network utilization. Indeed, a recent study shows that the ratio between the peak and the average rate is 3:1 for audio traffic, and as high as 15:1 for data traffic [88].

The main drawback of packet switching networks is that statistical multiplexing can lead to congestion. Network congestion happens when the arrival rate temporary exceeds the link capacity. In such a case, the network has to decide which traffic to drop, and which to transmit. In addition, end hosts are either expected to implement some form of congestion control, that is, to reduce their sending rates when they detect congestion in the network, or to avoid congestion by making sure that they do not send more traffic than the available capacity of the network.

Due to its superior flexibility and resource usage, the majority of today's networks are based on packet switching technologies. The most prominent packet switching architectures are Asynchronous Transfer Mode [3, 12], and Internet Protocol (IP) [22]. ATM uses fixed size packets called cells as the basic transmission unit, and was designed from the ground up to provide sophisticated services such as bandwidth and delay guarantees. In contrast, IP uses variable size packets, and supports only one basic service: best effort packet delivery, which does not provide any timeliness or reliability guarantees. Despite the advantages of ATM in terms of quality of service, during the last decade IP has emerged as the dominant architecture. For several technical and political reasons that try to explain this outcome see Tanenbaum [109].

As a result, our emphasis in this dissertation is on IP networks. While our SCORE/DPS techniques are applicable to packet switching networks in general, in this dissertation we examine them exclusively in the context of

IP. In the remainder of this chapter, we first present the Internet Protocol (IP) network model, which is the foundation of today's Internet, and then we consider some of the major proposals to provide better services in the Internet, and discuss their trade-offs.

2.2 IP Network Model

The main service provided by today's IP network is to deliver packets between any two nodes in the network with a "reasonable" probability of success. The key component that enables this service is the router. Each router has two or more interfaces that attach it to multiple networks. Routers forward each packet based on the destination address in the packet's header. For this purpose, each router maintains a table, called routing table, that maps every IP address to an interface attached to the router. Routing tables are constructed and maintained by the routing protocol. The routing protocol is implemented by a distributed algorithm whose main function is to let routers learn the reachability of any host in the Internet along a "good" path. In general, the term of "good" applies to the shortest[1] path to a node. Thus, ideally, a packet travels along the shortest path from source to destination.

2.2.1 Router Architecture

As noted in the previous section, a router consists of a set of input interfaces at which packets arrive, and a set of output interfaces, from which packets depart. The input and output interfaces are interconnected by a high speed fabric that allows packets to be transfered from inputs to outputs. The main parameter that characterizes the fabric is the *speedup*. The speedup is defined as the ratio between (a) the maximum transfer rate across the fabric from an input to an output interface, and (b) the capacity of an input (output) link.

As a packet traverses a router, the packet can be stored at input, at output, or at both the input and output interfaces. Based on where a router can store packets, routers are classified as *input* queueing, *output* queueing, or *input-output* queueing.

In an output-queueing router, when a packet arrives at the input, it is immediately transferred to the corresponding output. Since packets are enqueued and scheduled only at the outputs, this architecture is easy to analyze and understand. For this reason, most analytical studies assume an output-queueing router model.

On the downside, the output-queueing router architecture requires a speedup as high as n, where n is the number of inputs. The worst case scenario occurs when all inputs simultaneously receive packets for the same

---

[1]The most common metric used in today's Internet is the number of routers (hops) on the path.

output. Since inputs are bufferless, the output has to be able to *simultaneously* receive the n packets, hence the speedup of n. As the number of inputs in a modern router is quite large (e.g., it can exceed 32), building high-speed output-queueing routers is, in general, infeasible. That is why practically all of today's routers employ some sort of input-output queueing. By being able to buffer packets at the inputs, the speedup of the interconnection fabric can be significantly reduced. However this comes at a cost: complexity. Since only the output has complete knowledge of how packets are scheduled, complex distributed algorithms to control the packet transfer from inputs to outputs have to be implemented. Furthermore, this complexity makes the router behavior much more difficult to analyze.

In summary, while output-queueing routers are more tractable for analysis, the input and input-output queueing routers are more scalable and therefore easier to build. Fortunately, recent work has shown that a large class of algorithms implemented by an output queueing router can be emulated by an input-output queueing router which has an internal speedup of only 2 [21, 102]. Thus, at least in principle, it is possible to build scalable input-output queueing routers that can *emulate* the behavior of output queueing routers. *For this reason, in the remainder of this dissertation, we will assume an output queueing router architecture.*

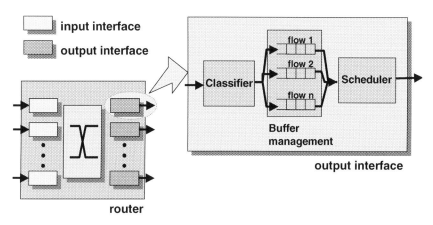

Fig. 2.1. The architecture of a router that provides per flow quality of service (QoS). Input interfaces use routing lookup or packet classification to select the appropriate output interface for each incoming packet, while output interfaces implement packet classification, buffer management, and packet scheduling. In today's best effort routers, neither input nor output interfaces implement packet classification.

Next, we discuss in more detail the output-queueing router architecture. Specifically, we present all the operations that a router needs to perform on the data and control paths in order to implement currently proposed solutions that aim to provide better services than the best effort service, such as Integrated Services and Differentiated Services.

2.2.2 Data Path

Data path represents the set of operations performed by routers on a data packet as the packet travels from source to destination. The main functions performed by routers on the data path are: (1) *routing lookup*, (2) *buffer management*, and (3) *packet scheduling*. Routing lookup identifies the output interface where to forward each incoming packet, based on the destination address in the packet's header. Buffer management and scheduling are concerned with managing router resources in case of congestion. In particular, when the buffer overflows, or when it exceeds some predefined threshold, the router has to decide what packet to drop. Similarly, when there is more than one packet in the buffer, the router has to decide what packet to transmit next. Usually, today's routers implement a simple drop-tail buffer management scheme, that is, when the buffer overflows, the packet at the tail of the queue is dropped. Packets are scheduled on a First-In-First-Out (FIFO) basis.

However, currently proposed solutions to provide more sophisticated services than the best effort service, such as per flow bandwidth and delay guarantees, require routers to perform a fourth function: (4) *packet classification*. Packet classification consists of mapping each incoming packet to the flow it belongs to. We use the term *flow* to denote a subset of packets that travel from one node to another node in the network. Since both routing lookup and packet classification can be used to determine to which output interface a packet is forwarded, in the remainder of this section we refer to both these operations as *packet forwarding* operations. Figure 2.1 depicts the relationship between the four functions in an output-queueing router that performs per flow management. In the remainder of this section, we present these functions in more detail. Since currently proposed solutions to provide per flow service semantics require routers to maintain and manage per flow state, we will elaborate on the complexity of these routers.

Packet Forwarding: Routing Lookup and Packet Classification Packet forwarding is the main and the most complex function performed by today's routers on the data path. This function consists of forwarding each incoming packet to the corresponding output interface based on the fields in the packet header. Virtually all routers in today's Internet forward packets based on their destination addresses. The process of finding the appropriate output port based on the packet destination address is called *routing lookup*. However, to implement more sophisticated functionalities such as providing better services to selected customers or filtering out some categories of traffic to enter the network, routers may need to use additional fields in the packet headers to distinguish between different traffic classes. Examples of such fields are the source address to identify the incoming traffic of a selected customer, and the destination port number to identify the traffic of different applications. The process of finding the class to which the packet belongs to is called *packet classification*. Note that routing lookup is a particular case of packet classification in which packets are classified based on one field: the destination address. In the remainder of this section we discuss in more detail routing lookup and packet classification.

Routing Lookup   With routing lookup, each router maintains a table, called routing table, that maps each IP address to an output interface. At the minimum, each entry in the routing table consists of two fields. The first field contains an address prefix, and the second field contains the identifier of an output interface. The address prefix specifies the range of all IP addresses that share the same prefix. Upon a packet arrival, the router searches its routing table for the longest prefix that matches the packet's destination address, and then forwards the packet to the output interface specified by the second field in the same entry. Thus, routing lookup consists of a search operation that retrieves the longest prefix match.

To minimize the size of the routing table, IP addresses are assigned in blocks based on their prefixes [41]. As a result, the size of the largest routing tables today is about 70,000 entries [122], which is three orders of magnitude smaller than the total number of hosts, which is about 72 million [89].

Traditional algorithms to implement the routing lookup are based on Patricia tries [72]. In the simplest form, Patricia tries are binary trees in which each node represents a binary string that encodes the path from the tree's root to that node. As an example, consider such a tree in which all left branches are labeled by 0, and all right branches are labeled by 1. Then, string 010 corresponds to the node that can be reached by walking down the tree from the root, first along the left branch, then along the right branch, and finally along the left branch. In the case of Patricia tries used to implement routing lookup, each leaf node represents an address prefix. Since the height of the tree is bounded by the address size s, the worst case time complexity of the lookup operation is $O(s)$. However, recent developments have significantly reduced this complexity. In particular, Waldvogel et al. [115] proposes a routing lookup algorithm that scales with the logarithm of the address size, while Degermark et al. [30] proposes a routing lookup algorithm tuned for IPv4 that takes less than 100 instructions on an Alpha processor, and uses only up to eight memory references. Furthermore, by using a hardware implementation, Gupta et al. [50] proposes a pipelined architecture that can perform a routing lookup every memory cycle. However, these improvements do not come for free. The complexity of updating the routing table in these algorithms is much higher than in the case of the algorithms based on Patricia tries. Nevertheless, this tradeoff is justified by the fact that, in practice, updates are much less frequent than lookups.

In summary, today it is possible to perform a routing lookup at the line speed, that is, without slowing down a router that otherwise performs only packet queuing and dequeuing. Furthermore, this is expected to remain true in the foreseeable future. Even if the Internet continues to expand at its current rate, due to address aggregation, the routing tables are likely to remain relatively small. Assuming that the current ratio between the number of hosts and the size of the routing tables will not change, and that, as predicted, the number of hosts will reach one billion by 2008 [88], we expect that routing table size will increase by a factor of about 16 over the next eight years. While this increase might seem considerable, it should be more than compensated for by the increase in computer processing power and memory capacity. Indeed, according to Moore's law, during the same time span the semiconductor performances are expected to improve 40 times.

Packet Classification   Current proposed solutions to provide Quality of Service (QoS) such as bandwidth and delay guarantees, require routers to maintain and manage per flow state. That is, upon a packet arrival, the router has to classify it to the class or flow the packet belongs to. A class is usually defined by a filter. A filter consists of a set of partially specified fields that define a region in the packet space. Common fields used for packet classification are source and destination IP addresses, source and destination port numbers, and the protocol type. An example of filter is ($src\_addr =$ 128.16.120.x, $dst\_addr =$ 234.16.120.x, $dst\_port =$ x, $src_port =$1000-1200, $proto\_type =$ x), where x stands for "don't care". This filter represents the entire traffic going from subnet 123.16.120.x to subnet 234.16.120.x with the destination port in the range 1000-1200. As an example, the packet identified by ($src\_addr =$ 123.16.120.12, $dst\_addr =$ 234.16.120.2, $dst\_port =$ 21, $src_port =$1080, $proto\_type =$ TCP) belongs to this class, while a packet sent by a host with the IP address 15.14.51.12 does not. It is worth noting that routing is just a particular case of packet classification, in which each filter is specified by only one field: dst_addr.

It should come as no surprise that the classification problem is inherently difficult. Current solutions [51, 66, 96, 97] work well only for a relatively small number of classes, i.e., no more than several thousand. This is because, as noted by Gupta and McKeown [51], the packet classification problem is similar to the point location problem in the domain of computation geometry. Given a point in an F dimensional space, this problem asks to find the enclosing region among a set of regions. In the case of non-overlapping regions, the best bounds for n regions in an F dimensional space are $O(\log n)$ in time and $O(n^F)$ in space, or, alternatively, $O(\log^{F-1} n)$ in time and $O(n)$ in space. This suggests a clear trade-off between space and time complexities. It also suggests that it is very hard to simultaneously achieve both speed and efficient memory usage. Worse yet, the packet classification problem is even more difficult than the traditional point location problem as it allows class (region) overlapping.

Buffer Management IP routers are based on a store-and-forward architecture, i.e., when a packet arrives at a router, the packet is first stored in a buffer, and then forwarded. Since, in practice, buffers are finite, the routers have to cope with the possibility of packet loss. Even with infinite buffer capacity, there might be the need to drop packets, as some congestion control schemes, such as TCP, rely on packet loss to detect network congestion.

Any buffer management scheme has to answer two questions: (1) *when* is a packet dropped?, and (2) *which* packet is dropped? In addition, *per flow* buffer management schemes have to answer a third question: (3) *which* queue to drop from? Examples of policies that answer the first question are: drop a packet when the buffer overflows (e.g., drop-tail), or when the average occupancy of the buffer exceeds some threshold. Examples of policies that answer the second question are: drop the last packet in the queue, the first packet in the queue, or a random packet. Finally, an example of policy that answers the last question is to drop a packet from the longest queue.

While simple network services can be implemented by using a single queue which is shared by all flows, solutions that provide more powerful services such as per flow bandwidth and delay guarantees require routers to maintain and manage a separate queue for each flow. In this case, the most expensive operation is usually to answer question (3), that is, to choose the queue to drop from. As an example, an algorithm that implements a policy that drops the packet from the longest queue has $O(\log n)$ complexity, where n is the number of non-empty queues. However, in practice, this complexity can be significantly reduced by grouping the queues that have the same size, or by approximating the algorithm [108].

Packet Scheduling The job of the packet scheduler is to decide what packet to transmit, if any, when the output link becomes idle. In routers that maintain per flow state this is accomplished in two steps: (1) select a flow that has a packet to send, and (2) transmit a packet from the flow's queue.

Packet scheduling disciplines are classified into two broad categories: *work conserving* and *non-work conserving*. In a work conserving discipline, the output link is busy as long as there is at least one packet in the system destined for that output. In contrast, in a non-work conserving discipline it is possible for an output link to be idle, despite the fact that there are packets in the system destined for that output. Virtually all routers in today's Internet are work-conserving, and implement a simple FIFO scheduling discipline. However, solutions to support better services than best effort, such as bandwidth and delay guarantees, require more sophisticated packet scheduling schemes. Examples of such schemes that are work conserving are: Static Priority [123], Weighted Round Robin [52], Virtual Clock [127], Weighted Fair Queueing [31], and Delay Earliest Deadline Due [124]. Similarly, examples of non-work conserving disciplines are: Stop-and-Go [44], Jitter-Virtual Clock [126], Hierarchical Round Robin [63], and Rate Controlled Static Priority [123].

Many of the simpler disciplines such as FIFO, Static Priority, and Weighted Round Robin can be easily implemented by constant time algorithms, i.e., algorithms that take $O(1)$ time to process each packet. In contrast, the more sophisticated scheduling disciplines such as Virtual Clock and Weighted Fair Queueing are significantly more complex to implement. In general, the algorithms to implement these disciplines associate with each flow a unique parameter that is used to select the flow to be served. Examples of such a parameter are the flow's priority, and the deadline of the packet at the head of the queue. Flow selection is usually implemented by selecting the flow with the largest or the smallest value. This can be accomplished by maintaining a priority queue data structure in which the time complexity of selecting a flow is $O(\log n)$, where n represents the number of flows in the queue.

Non-work-conserving disciplines, as well as some of the more complex work-conserving disciplines, may employ a second parameter. The purpose of the second parameter is to determine whether the flow with a non-empty queue is allowed to send or not. An example of such a parameter is the eligible time. The packet at the head of the queue can be transmitted only if its eligible time is smaller or equal to the system time. Obviously, the addition of a second parameter increases the implementation complexity. In many cases, the implementation is divided into two parts: a rate controller that stores packets until they become eligible, and a scheduler that selects the flow's packet to be transmitted based on the first parameter (e.g., deadline). Since the rate controller is usually implemented by constant time algorithms [10], the overall complexity of selecting a packet is generally dominated by the scheduling algorithm.

Once a flow is selected, one of its packets is transmitted – usually the packet at the head of the queue – and the parameter(s) associated with the flow are eventually updated.

2.2.3 Control Path

The *control path* consists of all functions and operations performed by the network to set up and maintain the state required by the data path. These functions are implemented by *routing* and *signaling* protocols.

Routing Protocol The purpose of routing protocols is to set up and maintain routing tables of all routers in a network. Routing protocols are implemented by distributed algorithms that try to learn the reachability of any host in the network. In the Internet, routing protocols are organized in a two level hierarchy.

At the higher level, the Internet consists of a large number of interconnected autonomous systems (ASs). An AS represents a distinct routing domain, which is usually administrated by a single organization such as a company or university. ASs are connected via gateways, which use *inter*-domain
routing protocols to exchange routing information about which hosts are
reachable by each AS. As a result, each gateway constructs a routing table
that maps each IP address to a neighbor AS that knows a path to that IP
address. The most common inter-domain routing protocol in use today is
Border Gateway Protocol (BGP) [86].

At the lower level within an AS, routers communicate with each other
using an *intra*-domain routing protocol. The purpose of these protocols is to
enable routers to exchange locally obtained information so that all routers
within an AS have coherent and up to date information needed to reach any
host within the AS. Examples of intra-domain routing protocols are Routing
Information Protocol (RIP) [54], and Open Shortest Path First (OSPF) [73].

The division of routing protocols into intra- and inter-domain is crucial
for the scalability of the Internet. On one hand, this allows the deployment
of sophisticated inter-routing protocols which can gather an accurate picture
of the host reachability within an AS. On the other hand, the inter-domain
routing protocols present a much coarser information about host reachability. Unlike intra-domain routing protocols that specify the path at the router
granularity, these protocols specify the path at the AS granularity. This trade-
off gives an organization maximum flexibility in managing its own resources,
without compromising routing scalability at the level of the entire Internet.
Some key factors affecting routing scalability, as well as some basic principles
of designing scalable routing protocols are presented by Yo [122].

In summary, as proved by the Internet's own existence, the hierarchical
routing architecture is both scalable and robust. However, it should be noted
that one of the main motivations behind these desirable properties is the
weak semantic of the best effort service. The best effort service does not
provide any reliability or timeliness guarantees. As long as a "reasonable"
number of packets reach their destinations, packet loss and packet reordering
are acceptable. As a result, route changes, route oscillations, or even router
failures do not necessary compromise the service. In contrast, with stronger
service semantics such as the guaranteed service, existing routing protocols
are not good enough. The next section discusses these issues in more detail.

Signaling Protocol To implement more sophisticated services such as per
flow delay and bandwidth guarantees, we need the ability to perform *admission control* and *route pinning*. The task of admission control is to reserve
enough resources on the path from source to destination in order to meet
the service requirements. In turn, route pinning makes sure that all packets
of the flow traverse the path on which resources have been reserved. Traditionally, these two functionalities are implemented by signalling protocols
such as Tenet Real-Time Channel Administration Protocol (RCAP) [8, 34],

or RSVP [128]. The complexity of signaling protocols is primary due to the difficulty of maintaining the state consistent in a distributed environment. In the remainder of this section we discus this issue in the context of both admission control and route pinning.

Admission Control Admission control makes sure that there are enough network resources on the path from source to destination to meet the service requirements, such as delay and bandwidth guarantees. To better understand the issues with admission control consider the following example. Assume host A requests bandwidth reservation for a flow that has destination B. One possible method to achieve this is to send a control message embedding the reservation request along the path from A to B. Upon receiving this message, each router along the path checks whether it has enough resources to accept the reservation. If it does, it allocates the required resources and then forwards the message. When host B receives this message, it sends back an acknowledgement to A. The reservation is considered successful if and only if all routers along the path have accepted it; otherwise the reservation is rejected. While simple, this procedure does not account for various failures such as packet loss and partial reservation failures. Partial reservation failures occur when only a subset of routers along the path accept the reservation. In this case, the protocol has to undo the reservation at the routers that have accepted it. To handle packet loss, when a router receives a reservation request message, the router has to be able to tell whether it is a duplicate of a message already processed or not. To handle partial reservation failures, a router needs to remember the decision made for the reservation request in a previous pass. For these reasons, all existing solutions maintain per flow reservation state, be it hard state as in ATM UNI [1], Tenet Real-Time Channel Administration Protocol (RCAP) [8, 34], or soft state as in RSVP [128]. However, maintaining *consistent* and *dynamic* state in a *distributed* environment is in itself a challenging problem. Fundamentally, this is because admission control assumes a transaction-like semantic, which is very difficult to achieve in a distributed system in the presence of message losses and arbitrary delays [4, 117].

Route Pinning Once a flow's reservation request is accepted, the source can start sending data packets. However, to meet the performance requirements negotiated during the admission control, we have to make sure that *all* packets of a flow traverse the same path. Otherwise, a packet can traverse a path that does not have enough resources, which will lead to service violation. The operation of binding a flow to a path (route) is called *route pinning*. Whenever the underlying routing protocol does not support route pinning, this functionality can be provided by the signalling protocols together with the admission control. For example, in RSVP, when a node accepts a reservation, it also stores the next hop on the flow's path in its database. Since these protocols maintain per flow state, augmenting this state to store the next hop does not increase their complexity.

Alternatively, route pinning can be separated from admission control. One example is ATM [12] whose routing protocol natively supports route pinning. Another example is Multi-Protocol Label Switching (MPLS), recently proposed to perform traffic engineering in the Internet [17]. In both ATM and MPLS, the main idea is to perform routing based on identifiers that have *local* meaning, instead of identifiers that have *global* meaning such as IP addresses. Each router maintains a table which maps each local identifier (label) to an output interface. Each packet carries a label that specifies how the packet is to be routed at the next hop. Before forwarding a packet, a router replaces the existing label with a new label that is used by the next hop to route the packet. Note that this requires a router to also store the labels used by its neighbors. Besides route pinning, one other major advantage of routing based on labels, instead of IP addresses, is performance. Instead of searching for the longest prefix match, we only have to search for an exact match, which is much faster to implement.

On the downside, these routing schemes need a special protocol to distribute and maintain labels consistent. While in this case routers do not need to maintain per flow state, they still need to maintain per path state. However, in practice, the number of paths that can traverse a core router can be still quite large. In the worst case, this number increases with the square of the number of edge nodes. Thus, in the case of an AS that has hundreds of edge nodes, this number can be on the order of hundred of thousands. Finally, label distribution protocols have to address the same challenges as other distributed algorithms that need to maintain state consistent in the presence of link and router failures, such as Tenet RCAP and RSVP.

2.2.4 Discussion

Among all the operations performed by routers on the data path, packet classification is arguably the most complex. As discussed in Section 2.2.2, algorithms to solve this problem require at least $O(\log n)$ time and $O(n^F)$ space, or, alternatively, at least $O(\log^{F-1})$ time and $O(n)$ space, where n represents the number of classes, and F represents the number of fields in a filter.

In contrast, most buffer management and packet scheduling algorithms have $O(n)$ space complexity and $O(\log n)$ time complexity. By trading resource utilization for speed, we can further reduce the time complexity to $O(\log \log n)$ or even $O(1)$. For example, [98] proposes an implementation of Weighted Fair Queueing with $O(1)$ time complexity.

The important point to note here is that our DPS technique trivially eliminates the most complex operation performed by core routers on the data path: packet classification. This is because, with DPS, the state required to process packets is carried by the packets themselves, instead of being maintained by core routers (see Section 1.1). Consequently core routers do not need to perform any packet classification.

On the control path the most complex operation is arguably the admission control for which current solutions require routers to maintain per flow state. The main difficulty is to maintain the consistency of the distributed state in the presence of packet losses, arbitrary packet delays, and router failures.

Again, the main benefit of using DPS is that by eliminating the need for core routers to maintain per flow state, we trivially eliminate the need of maintaining this state consistent.

2.3 Network Service Taxonomy

In this section we present a general taxonomy of services in a packet switching network, and then use this taxonomy to describe the traditional best effort service, and the recently proposed services to enhance today's Internet. We then describe and compare the existing solutions to implement these services.

The primary goal of a network is to provide services to its end-hosts. Services are usually classified along two axes: (a) the granularity of the network abstraction to which the service applies, and (b) the "quality" of the service.

As the name suggests, packet switching networks are centered around the *packet* abstraction. A packet represents the smallest piece of information that can be routed through the network. At a higher level of granularity, we have the concept of a *flow*. A flow represents a subset of packets that travel between two nodes in the network. If these nodes are routers, we will also use the terminology of *macro-flow*. An example of a flow is the traffic of a TCP connection, while an example of a macro-flow is the traffic between two sub-networks. At an even higher level of abstraction, we have traffic *aggregates* over multiple destinations or sources. Examples of traffic aggregates are the entire web traffic of a user, or the entire outgoing/incoming traffic of an organization.

Along the second axis, a service is described by a set of properties that can be either *qualitative* or *quantitative*. Examples of qualitative properties are *reliability* and *isolation*. Isolation refers to the ability of the network to protect the traffic of a flow against malicious sources that may flood the network. Quantitative properties are described in terms of performance parameters such as bandwidth, delay, delay jitter and loss probability. Usually, these parameters are reported on an end-to-end basis. For example, the delay represents the total time it takes a packet to travel from source to its destination. Similarly, the delay jitter represents the maximum difference between the maximum and the minimum end-to-end delays experienced by any two packets of a flow. Note that the two quantitative and qualitative properties are not necessary orthogonal. For example, a service that guarantees a zero loss probability is trivially a reliable service.

Quantitative services can be further classified into *absolute* and *relative* services. Absolute services specify precise quantities that bound the service performance parameters such as worst case bandwidth or delay. In contrast, relative services specify the relative difference or ratio between the performance parameters. Examples of absolute services are: "flow A is guaranteed a bandwidth of 2 Mbps", and "the loss probability of flow A is less than 10^{-6}". Examples of relative services are: "flow A has twice the bandwidth of flow B", and "flow A has a packet loss twice as small as flow B".

Next, we discuss some of the most prominent services proposed in the context of the Internet: (1) the best effort service, (2) flow protection to provide network support for congestion control, (3) Integrated Services, and (4) Differentiated Services. Table 2.1 shows a taxonomy of these services.

Service		Network Abstraction	Service Description
Best effort		packet	connectivity
Flow Protection		flow	protect well-behaved flows against ill-behaved ones
Intserv	Guaranteed	flow	bandwidth and delay guarantees
	Controlled-Load	flow	"weak" bandwidth guarantees
Diffserv	Premium	macro-flow	bandwidth guarantees
	Assured	traffic aggregate over multiple receivers/sources	"weak" bandwidth guarantees

Table 2.1. A taxonomy of services in IP networks.

2.3.1 Best Effort Service

Today's Internet provides one simple service: the best effort service. This is fundamentally a connectivity service which allows any two hosts in the Internet to communicate by exchanging packets. As the name suggests, this service does not make any promise of whether a packet is actually delivered to the destination, or whether the packets are delivered in order or not. Such a minimalist service requires little support from routers. In general, routers just forward packets on a First-In First-Out (FIFO) basis. Thus, excepting the routing state, which is highly aggregated, a router does not need to maintain and manage any fine grained state about traffic. This simple architecture has several desirable properties:

Scalability Since the only state maintained by routers is the routing state, today's Internet architecture is highly scalable. In particular, address aggregation allows routers to maintain little state as compared to the number of hosts in the network. For example, a typical router today stores less than

70,000 entries [122] which is several orders of magnitude lower than the number of hosts in the Internet, which is around 72 million [89].

Robustness One of the most important goals in designing the Internet was robustness [22]. In particular, the requirement was that two end-hosts should be able to communicate despite router and link failures, and/or network reconfiguration. The only case in which two hosts can no longer communicate is when the network between the two hosts is partitioned. The fact that the state of a flow is maintained only by end-hosts and not by routers makes it significantly easier to ensure robustness, as router failures do not compromise the flow state. Had the flow state been kept by routers, complex algorithms to replicate and restore this state would be needed to handle failures. Furthermore, such algorithms would be able to provide protection against failures only if the number of routers failing is smaller than the number of replicas that failed.

There is one caveat with respect to the Internet robustness though. It can be argued that, to a large extent, today's Internet is robust mainly because it provides a weak service semantic. Indeed, as long as the "majority" of packets still reach their destination, router or link failures do not compromise the service. In contrast, it is fundamentally more difficult to achieve robustness in the case of a strong semantic service such as the guaranteed service. In this case, a router or link failure can easily compromise the service. Note that even if back-up paths were used to restore the service, time sensitive parameters such as delay may still be affected during the recovery process.

Performance The simplicity of the router design allows efficient implementation at very high speeds. Usually, these routers implement the FIFO scheduling discipline and drop-tail buffer management, which are both constant-time operations.

2.3.2 Flow Protection: Network Support for Congestion Control

Because of their reliance on statistical multiplexing, data networks such as the Internet must provide mechanisms to control congestion. The current Internet relies on end-to-end congestion control mechanisms in which senders reduce their transmission rates whenever they detect congestion in the network. The most widely utilized form of congestion control is the additive-increase/multiplicative-decrease scheme implemented by TCP [57, 83], a scheme which has proven to be highly successful in preventing congestion collapse[2]. However, the viability of this approach depends on one fundamental assumption: all end-hosts *cooperate* by implementing *equivalent* congestion control algorithms.

---

[2]Congestion collapse occurs when sources increase their sending rates when they experience losses, in the hope that more of their packets will get through. Eventually, this will lead to a further increase in the packet loss, and result in consistent buffer overflow at the congested routers.

While this was a reasonable assumption when the Internet was primarily used by the research community, and the vast majority of traffic was TCP based, this is no longer true today. The emergence of new multimedia applications, such IP telephony, audio and video streaming, which use more aggressive UDP based protocols, negatively affects the still predominant TCP traffic. Although there are considerable ongoing efforts to develop protocols for the new applications that are TCP *friendly* [6, 84, 85] – protocols that implement TCP like congestion control algorithms – these efforts fail to address the fundamental problem: *in an economic environment cooperation is not always optimal.* In particular, in case of congestion, the natural incentive of a sender is to send more traffic in the hope that it will force other senders to back-off, and as a result it will be able to use the extra bandwidth. This incentive translates into a positive feed-back behavior, i.e., the more packets that are dropped in the network, the more packets the user sends, which can ultimately lead to congestion collapse. It is interesting to note that this problem resembles the "tragedies of commons" problem, well known in the economic literature [53].

Two approaches were proposed to address this problem: (1) *flow identification* and (2) *fair bandwidth allocation.* Both of these approaches require changes in the routers. In the following sections, we discuss these approaches in more detail.

Identification Approach The main idea of this approach, advocated by Floyd and Fall [36], is to identify and then punish the flows that are *ill-behaved.* In short, routers employ a set of tests to identify ill-behaved flows. When a flow is identified as being ill-behaved, it is punished by preferentially having its packets dropped until its allocated bandwidth becomes smaller than the bandwidth allocated to a well-behaved flow. In this way, the punishment creates the incentive for end-hosts to send well-behaved traffic. The obvious question is how to identify an ill-behaved flow.

To answer this question, Floyd and Fall [36] propose a suite of tests, which try to detect whether a flow is *TCP friendly* or not, i.e., whether the behavior of a flow is consistent to the behavior of a TCP flow under similar conditions. In particular, these tests estimate the round-trip time (RTT) and the packet dropping probability, and then check whether the throughput of a flow and its dynamics are consistent to those of a TCP flow having the same RTT and experiencing the same packet dropping probability.

While this approach can be efficiently implemented, it has two significant drawbacks. First, these tests are generally inaccurate as they are based on parameters that are very hard to estimate. For example, it is very difficult if not impossible to accurately estimate the RTT of an arbitrary flow based only on the local information available at the router, as assumed by Floyd and Fall [36]. [3] Because of this, current proposals simply assume that the

---

[3]While a possible solution would be to have the end-hosts sending the estimated RTT to routers along the flow's path, there are two problems with this approach.

RTT is twice the propagation delay on the outgoing link. Clearly, depending on the router position on the path of the flow, this procedure can lead to major under-estimations, negatively impacting the overall accuracy of these tests.

Second, this approach makes the implicit assumption that *all* existing and future congestion protocol algorithms are going to be TCP friendly. From an architectural standpoint, this assumption considerably reduces the freedom of designing and building new protocols. This can have significant implications, as the freedom allowed by the original datagram service, one of the key properties that has contributed to the success of the Internet, is lost.

Allocation Approach In this approach routers employ special mechanisms that allocate bandwidth in a fair manner. Fair bandwidth allocation protects well-behaved flows from ill-behaved ones, and is typically achieved by using per-flow queueing mechanisms such as Fair Queueing [31, 79] and its many variants [10, 45, 94].

Unlike the identification approach, the allocation approach allows various congestion policies to coexist. This is because no matter how much traffic a source will send in the network, it is not going to get more than its fair allocation. Unfortunately, this flexibility does not come for free. Fair allocation mechanisms are complex to implement, as they inherently require routers to maintain state and perform operations on a per flow basis. In contrast, with the identification approach, routers need to maintain state only for the flows which are punished, i.e., the ill-behaved flows.

2.3.3 Integrated Services

As new applications such as IP telephony, video-conferencing, audio and video streaming, and distributed games are deployed in the Internet, services more sophisticated than best effort are needed. Unlike previous applications such as file transfer, these new applications have much stricter timeliness and bandwidth requirements. For example, to enable natural interaction, the end-to-end delay needs to be below human perception. Previous studies concluded that for natural hearing this delay should be around 100 ms [64]. Since in a global network the propagation delay alone is about 100 ms, meeting such tight delay requirements is a very challenging task [7].

To support these new applications, IETF has proposed a new service model called Integrated Services or Intserv [82]. Intserv uses flow abstraction. Two services were defined within the Intserv framework: *Guaranteed* and *Controlled-Load* services.

First it requires that changes be made to the end-hosts, and second, there is the question of whether a router can trust this information.

Guaranteed Service Guaranteed service is the strongest semantic service proposed in the context of the Internet so far [93]. Guaranteed service has the ability to provide per flow bandwidth and delay guarantees. In particular, a flow can be guaranteed a minimum bandwidth, and, given the arrival process of the flow, a maximum end-to-end delay. This way, Guaranteed service provides ideal support for real-time applications such as IP telephony.

However, this comes at the cost of a significant increase in complexity: current solutions require routers to maintain and manage per flow state on both data and control paths. On the data path, a router has to perform per flow classification, buffer management and scheduling. On the control path, routers have to maintain per flow forwarding state and perform per flow admission control. During the admission control, each router on the flow's path reserves network resources, such as the link capacity and buffer space, to make sure that the flow's bandwidth and delay requirements are met.

Controlled-Load Service For applications that do not require strict service guarantees, IETF has proposed a weaker semantic service within the Intserv framework: the Controlled-Load service. As defined by Wroclawski [121], the Controlled-Load service *"tightly approximates the behavior visible to applications receiving best effort service *under unloaded conditions* from the same series of network elements"*. More precisely, the Controlled-Load service ensures that (1) the packet loss is not significantly larger than the basic error rate of the transmission medium, and (2) the end-to-end delay experienced by a very large percentage of packets does not greatly exceed the end-to-end propagation delay. The Controlled-Load service is intended to provide better support for a broad class of applications that have been developed for use in today's Internet. Among the applications that fall into this class are the "adaptive and real-time applications" such as video and audio streaming.

While the Controlled-Load service still requires routers to perform per flow admission control on the control path, and packet classification, buffer management, and scheduling on the data path, some of these operations can be significantly simplified. For example, the scheduling can be implemented by a simply weighted round robin discipline, which has $O(1)$ time complexity. Thus, the Controlled-Load trades a lower quality of service for a simpler implementation.

In summary, although Intserv provides much more powerful and flexible services than today's Internet – services that would answer the needs of the new emerging applications – concerns with respect to its complexity and scalability have hampered its adoption. In fact, except in small test-beds, Intserv solutions have yet to be deployed.

2.3.4 Differentiated Services

To alleviate the scalability problems that have plagued Intserv, recently a new service model, called Differentiated Services (Diffserv), has been proposed [13,

75]. The Diffserv architecture differentiates between edge and core routers. Edge routers maintain per flow or per aggregate state. Core routers maintain state only for a very small number of traffic classes; they do not maintain any fine grained state about the traffic. Each packet carries in its header a six bit field, called the Differentiated Service (DS) field, which specifies the class to which the packet belongs. The DS field is initialized by the ingress router upon the packet arrival. In turn, core routers use the DS field to classify and process the packets. Since the number of classes at a core router is very small, packet processing can be very efficiently implemented. This makes the Diffserv architecture highly scalable.

Two services were proposed in the context of the Diffserv architecture: *Assured* and *Premium* services.

Assured Service The Assured service [24, 55] is a large granularity service, that is, the service is associated with the aggregate traffic of a customer from/to multiple hosts. The service contract between a customer and the Diffserv network or ISP is called the *service profile*. A service profile is usually defined in terms of absolute bandwidth and relative loss. As an example, an ISP can provide two service levels (classes): silver and gold, where the gold service has the lowest loss probability. A possible service profile would offer transmission of 10 Mbps of customer's web traffic by using the silver service.

In the Assured service model, ingress routers perform three functions. They (a) monitor the aggregate traffic from each user to make sure that no user exceeds its traffic profile, (b) downgrade the user's traffic to a lower service level if the user exceeds its profile, and (c) initialize the DS field in the packet headers with the code-point associated to the service. Thus, ingress routers need to keep state for each profile or user. In contrast, core routers do not need to keep such state, as their function reduces to process the packets based on the code-points carried by the packets.

While the fixed bandwidth profile makes the Assured service very compelling, it also makes it very challenging to implement. This is due to a fundamental conflict between maximizing resource utilization and achieving high service assurance. Since a service profile does not specify how the traffic is distributed through the network, the network has to make conservative assumptions to achieve high service assurance. At the limit, to guarantee zero loss, the network has to assume that the entire assured traffic traverses the slowest link in the network! Clearly, such an assumption leads to a very low resource utilization, which can be unacceptable.

An alternate approach is to define service profiles in relative rather than absolute terms. Such an example is the User-Share Differentiation (USD) approach [116]. With USD each user is assigned a share (weight) that specifies the relative fraction of the capacity that a user is entitled to receive on each link in the network. This is equivalent to a network in which the capacity of each link is allocated by a Weighted Fair Queueing algorithm. The problem with such an approach is that the core routers need to maintain per user state, which can negate the scalability advantage of the Diffserv architecture. In addition, with USD, there is little correlation between the share of a user and the aggregate throughput it will receive. For example, two users that are assigned the same share can see drastically different aggregate throughputs. A user that has traffic for many destinations (thus traverse many different paths) can potentially receive much higher aggregate throughput than a user that has traffic for only a few destinations.

Premium Service Unlike the Assured service which can be associated with an aggregated traffic to/from multiple hosts, Premium service provides the equivalent of a dedicated link of fixed bandwidth between two edge routers [60]. To implement this service, the network has to perform admission control. The current proposals assume a centralized architecture: each domain is associated with a database, called Bandwidth Broker (BB), that has complete knowledge about the entire domain. To set up a flow across a domain, the domain's BB checks first whether there are enough resources between the two end points of the flow across the domain. If yes, the request is granted and the BB's database is updated accordingly.

On the data-path, ingress routers perform two functions. They (a) shape the traffic associated to a service profile, that is, make sure that the traffic does not exceed the profile by delaying the excess packets[4], and (b) insert the Premium service code-point in the DS-files. In turn, core routers forward the premium traffic with high priority.

As a result, the Premium service can provide effective support for real-time traffic. A natural question to ask is what is the difference between the Premium service and the Guaranteed service proposed by Intserv. Though at the surface they are quite similar, there are two important differences between them.

First, while the Guaranteed service can provide both per flow bandwidth and delay differentiation, the Premium service can provide only per flow bandwidth differentiation. This is because core routers do not differentiate between premium packets on a per flow basis – all premium packets are simply processed in a FIFO order. Thus, the only possibility to meet different delay requirements for different flows is to guarantee the smallest delay required by any flow to *all* flows. Unfortunately, this can result in very low resource utilization for the premium traffic. In particular, as shown by Stoica and Zhang [105], even if the fraction that can be allocated to premium traffic on every link in the network is very low (e.g., 10%), the worst case queueing delay across a large network (e.g., 15 routers) can be relatively large (e.g., 240 ms). In contrast, Intserv can achieve both higher resource utilization and tighter delay bounds, by better matching flow requirements to resource usage.

Second, the centralized bandwidth broker architecture proposed to perform admission control in the case of the Premium service is adequate only for

---

[4]Note that this is different from the Assured service, where the excess traffic is let into the network, but its priority is downgraded.

coarse grained flows that are active over long time scales. In contrast, because the Guaranteed service uses a distributed admission control architecture, it can support fine grained reservations over small time scales.

The price paid by the Guaranteed service is again complexity. Unlike the Premium service, the Guaranteed service requires routers to maintain per flow state on both the data and the control paths.

2.4 Summary

In the first part of this chapter, we have discussed the IP network model. In particular we have presented the router architecture and discussed the implementation complexities of both the data and control paths.

In the second part of this chapter, we have presented the best-known proposals to improve the best effort service in today's Internet: (a) flow protection to provide effective support for congestion control, (b) Integrated Services (Intserv) model, and (c) Differentiated Services (Diffserv) model. Of all these models, only Diffserv admits a known scalable implementation, as core routers are not required to maintain any per flow state. However, to achieve this, Diffserv makes significant compromises. In particular, the Assured service cannot achieve simultaneously high service assurance and high resource utilization. Similarly, the Premium service cannot provide per flow delay differentiation, and it is not adequate for fine grained and short term reservations.

In this dissertation we address these shortcomings by developing a novel solution that can implement *all* of the above per flow services (i.e., flow protection, guaranteed and controlled-load services) in the Internet without compromising its scalability. In the next chapter, we present an overview of our solution.

3 Overview

The main contribution of this dissertation is to provide the first solution that makes it possible to implement services as powerful and as flexible as the ones implemented by a stateful network using a stateless core network architecture. In this chapter, we give an overview of our solution and present a perspective of how this solution compares to the two most prominent solutions proposed in the literature to provide better services in the Internet: Integrated Services and Differentiated Services.

The chapter is organized as follows. Section 3.1 describes the main components of our solution and uses three examples to illustrate the ability of our key technique, called Dynamic Packet State (DPS), to provide per flow functionalities in a stateless core network. Section 3.2 briefly describes our implementation prototype, and gives a simple example to illustrate the capabilities of our implementation. Section 3.3 presents a comparison between our solution and the two main network architectures proposed by Internet Engineering Task Force (IETF) to provide more sophisticated services in the Internet: Integrated Services and Differentiated Services. Finally, Section 3.4 summarizes our findings.

3.1 Solution Overview

This section presents the three main components of our solution. Section 3.1.1 defines the Stateless Core (SCORE) network architecture, which represents the basic building block of our solution. Section 3.1.2 presents a novel approach that allows us to emulate/approximate the service provided by a stateful network with a SCORE network. Section 3.1.3 describes the key technique we use to implement this approach: Dynamic Packet State (DPS). To illustrate this technique we sketch how it can be used to implement three per flow mechanisms in a SCORE network: (1) approximate Fair Queueing scheduling discipline, (2) provide per flow admission control, and (3) perform route pinning.

I. Stoica: Stateless Core, LNCS 2979, pp. 35-51, 2004.

3.1.1 The Stateless Core (SCORE) Network Architecture

The basic building block of our solution is called Stateless Core (SCORE). Similar to a Diffserv domain, a SCORE domain is a *contiguous* and *trusted* region of network in which only edge routers maintain per flow state, while core routers maintain no per flow state (see Figure 1.1(b)). Since edge routers usually run at much lower speeds and handle fewer flows than the core routers, this architecture is highly scalable.

3.1.2 The "State-Elimination" Approach

Our ultimate goal is to provide better services in today's Internet without compromising its scalability and robustness. To achieve this goal, we propose a two step approach, called "State-Elimination" approach (see Figure 1.1). In the first step we define a *reference* stateful network that provides the desired service. In the second step, we try to approximate or, if possible, to emulate the service provided by the reference stateful network in a SCORE network. In this way we are able to provide services as powerful and as flexible as the ones implemented by stateful networks in a mostly stateless network, i.e., in a SCORE network. In Chapters 4, 5 and 6, we illustrate this approach by considering three of the most important services proposed to enhance today's Internet: *flow protection*, *guaranteed services*, and *relative service differentiation*.

We note that similar approaches have been proposed in the literature to approximate the functionality of an idealized router that implements a bit-by-bit round-robin scheduling discipline with a stateful router that forwards traffic on a per packet basis [10, 31, 45, 79]. However, our approach differs from these approaches in two significant aspects. First, the state-elimination approach is concerned with emulating the functionality of an *entire* network, rather than of a single router. Second, unlike previous approaches that aim to approximate an idealized system with a stateful system, our goal is to approximate the functionality of a stateful system with a stateless core system.

3.1.3 The Dynamic Packet State (DPS) Technique

DPS is the key technique that allows us to implement the above services in a SCORE network. The main idea behind DPS is very simple: *instead of having routers install and maintain per flow state, have packets carry the per flow state.* This state is inserted by ingress routers, which maintain per flow state. In turn, a core router processes each incoming packet based on (1) the state carried in the packet's header, and (2) the router's internal state. Before forwarding the packet to the next hop, the core router updates both its internal state and the state in the packet's header (see Figure 1.2). By using DPS to coordinate actions of edge and core routers along the path traversed by a flow, distributed algorithms can be designed to approximate the behavior of a broad class of stateful networks using networks in which core routers do not maintain per flow state.

To give an intuition of how the DPS technique is working, next we present three examples: (1) approximates the Fair Queueing algorithm, (2) estimates the aggregate reservation for admission control purposes, and (3) binds a flow to a particular path (i.e., perform route-pinning).

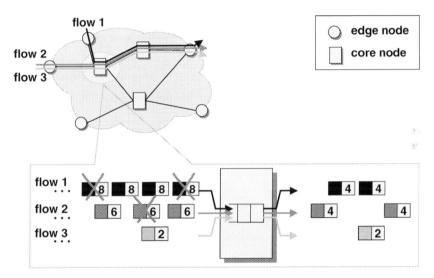

Fig. 3.1. Example illustrating the CSFQ algorithm at a core router. An output link with a capacity of 10 is shared by three flows with arrival rates of 8, 6, and 2, respectively. The fair rate of the output link in this case is $\alpha = 4$. Each arrival packet carries in its header the rate of the flow it belongs to. According to Eq. (3.3) the dropping probability for flow 1 is 0.5, while for flow 2 it is 0.33. Dropped packets are indicated by crosses. Before forwarding a packet, its header is updated to reflect the change in the flow's rate due to packet dropping (see the packets at the right-hand side of the router).

Example 1: Fair Bandwidth Allocation Flow protection is one of the most desirable enhancements of today's best effort service. Flow protection allows diverse end-to-end congestion control schemes to seamlessly coexist in the Internet, and protect well behaved traffic against malicious or ill behaved traffic. The solution of choice to achieve flow protection is to have routers implement *fair bandwidth allocation* [31]. In an idealized system in which a router can provide services at the bit granularity, fair bandwidth allocation can be achieved by using a bit-by-bit round robin discipline.

For clarity, consider three flows with the arrival rates of 8, 6, and 2 bits per second (bps), respectively, that share a 10 bps link. Assume that the traffic of each flow arrives one bit at a time, and it is periodic with period $1/r$, where

r is the flow rate. Thus, during one second, exactly 16 bits are received, and exactly 10 bits can be transmitted. During each round, the scheduler transmits exactly one bit form every flow that has a packet to send. Since in the worst case, flow 3 is visited once every 3/10 sec, and it has an arrival rate of only one bit every 0.5 sec, it follows that all of its traffic is served. This leaves the other two flows to share the rest of 8 bps of the link capacity. Since arrival rates of both flows 1 and 2 are larger than half of the remaining capacity, each flow will receive half of it, i.e., 4 bps. As a result, under the bit-by-bit round robin discipline, the three flows are allocated bandwidth of 4, 4, and 2 bps, respectively. The maximum rate allocated to a flow on a congestion link is called *fair rate*. In this example the fair rate is 4.

In general, given n flows that traverse a congested link of capacity, C, the fair rate α is defined such that

$$\sum_{i=1}^{n} \min(r_i, \alpha) = C, \tag{3.1}$$

where r_i represents the arrival rate of flow i. By applying this formula to the previous example, we have $\min(8, \alpha) + \min(6, \alpha) + \min(2, \alpha) = 10$, which gives us $\alpha = 4$. If the link is not congested, that is, if $\sum_{i=1}^{n} r_i < C$, the fair rate, α, is by convention defined as being the maximum among all arrival rates.

Thus, with the bib-by-bit round robin, the service rate allocated to a flow, i, with the arrival rate, r_i, is

$$\min(r_i, \alpha). \tag{3.2}$$

The first algorithm to approximate the bit-by-bit round robin in a packet system was proposed by Demers et al. [31], and it is called Fair Queueing. Eq. (3.2) directly illustrates the protection property of Fair Queueing, that is, a flow with enough demand is guaranteed to receive its fair rate α, *irrespective* of the behavior of the other flows. To put it in another way, a flow cannot deny service to other flows because no matter how much and what type of traffic it pumps into the network, it will not get more than α on the congested link.

While fair queueing can fully provide flow protection, it is more complex to implement than traditional FIFO queueing with drop-tail, which is the most widely implemented and deployed mechanism in routers today. For each packet that arrives at the router, the routers needs to classify the packet into a flow, update per flow state variables, and perform per flow scheduling.

Our goal is to eliminate this complexity from the network core by using a SCORE network architecture to *approximate* the functionality of a reference network in which every router performs Fair Queueing. In the following sections, we describe a DPS based algorithm, called Core-Stateless Fair Queueing (CSFQ) that achieves this goal.

The key idea of CSFQ is *to have each packet carry the rate estimate of the flow it belongs to.* Let $\widehat{r}_i$ denote the rate estimate carried by a packet of flow i. The rate estimate is computed by edge routers and then inserted in the packet header. Upon receiving the packet, a core router forwards it with the probability

$$p = \min\left(1, \frac{\alpha}{\widehat{r}_i}\right), \qquad (3.3)$$

and drops it with the probability $1 - p$.

It is easy to see that by forwarding each packet with the probability p, the router effectively allocates to flow, i, a rate $r_i \times p = \min(\widehat{r}_i, \alpha)$, which is exactly the rate the flow would receive under Fair Queueing (see Eq. (3.2)). If $p < 1$, the router also updates the packet label to α. This is to reflect the fact that when the flow's arrival rate is larger than α, the flow's rate after traversing the link drops to α (see Figure 3.1).

It is also easy to see that with CSFQ core routers do not require any per flow state. Upon packet arrival, a core router needs to compute only the dropping probability, p, which depends exclusively on the estimated rate carried by the packet, and the fair rate α that is locally computed by the router. (In Chapter 4, we show that computing α does not require per flow state either.)

Figure 3.1 shows an example in which three flows with incoming rates of 8, 6, and 2, respectively, share a link of capacity 10. Without going into details, we note that in this case $\alpha = 4$. Then, from Eq. (3.3), it follows that the forwarding probabilities of the three flows are 0.5, 0.66, and 1, respectively. As a result, on the average, one out of two packets of flow 1, one out of three packets of flow 2, and no packets of flow 3, are dropped. Note that before forwarding the packets of flows 1 and 2, the router updates the rate estimates in their headers to 4. This is to reflect the change of the flow rates as a result of packet dropping.

Example 2: Per Flow Admission Control In this example we consider the problem of performing per flow admission control. The role of the admission control is to check whether there are enough resources on the data path to grant a reservation request. For simplicity, we assume that admission control is limited to bandwidth. When a new flow makes a reservation request, each router on the path from source to destination checks whether it has enough bandwidth to accommodate the new flow. If all routers can accommodate the flow, then the reservation is granted.

It is easy to see that to decide whether a new reservation request rsv can be accepted or not, a router needs only to know the current aggregate reservation, R, on the output link, that is, how much bandwidth it has reserved so far. In particular, if the capacity of the output link is C, then the router can accept a reservation, rsv, as long as $rsv + R \leq C$. Unfortunately, it turns out that maintaining the aggregate reservation, R, in the presence

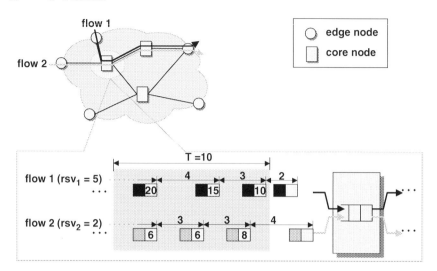

Fig. 3.2. Example illustrating the estimation of the aggregate reservation. Two flows with reservations of 5, and 2, respectively, share a common link. Ingress routers initialize the header of each packet according to Eq. (3.4). The aggregate reservation is estimated as the ratio between the sum of the values carried in the packets' headers during an averaging time interval of length T. In this case the estimated reservation is $65/10 = 6.5$.

of packet loss and partial reservation failures is not trivial. Intuitively, this is because the admission control needs to implement transaction-like semantics. A reservation is granted if and only if all routers along the path accept the reservation. If a router cannot accept a reservation, then all routers that have accepted the reservation have to roll back to the previous state, so they need to remember that state. Similarly, if a reservation request message is lost, and the request is resent, then a router has to remember whether it has received the original request, and if yes, whether the request was granted or denied. For all these reasons, the current proposed solutions for admission control such as RSVP [128] and ATM UNI [1] require routers to maintain per flow state.

In the remainder of this section we show that by using DPS it is possible to perform admission control in a SCORE network, that is, without core routers maintaining any per flow state.

At the basis of our scheme lies a simple observation: if all flows were sending at their reserved rates, then it is trivial to maintain the aggregate reservation R; each router only needs to measure the rate of the aggregate traffic. Consider the example in Figure 3.2, and assume that flow 1 has a reservation of 5 Kbps, and flow 2 has a reservation of 2 Kbps. If the two flows were sending exactly at their reserved rates, i.e., flow 1 at 5 Kbps, and flow 2 at 2 Kbps, the hi-lighted router (see Figure 3.2) can simply estimate the aggregate reservation of 7 Kbps, by measuring the rate of the aggregate arrival traffic.

The obvious problem with the above scheme is that most of the time flows do *not* send at their reserved rates. To address this problem, we associate a *virtual length* to each packet. The virtual length is such that if the lengths of all packets of a flow where equal to their virtual lengths, then the flow sends at its reserved rate. More precisely, the virtual length of a packet represents the amount of traffic that the flow was entitled to send according to its reserved rate since the previous packet has been transmitted. Let rsv_i denote the reservation of flow i, and let t_i^j and t_i^{j+1} denote the departure times of the j-th and $(j+1)$-th packets of flow i. Then the $(j+1)$-th packet will carry in its header a virtual length

$$vl_i^{j+1} = rsv_i \times (t_i^{j+1} - t_i^j). \qquad (3.4)$$

The virtual length of the first packet of the flow is simply the actual length of the packet. The virtual length is computed and inserted in the packet header by the ingress router upon the packet departure. In turn, core routers use the packet virtual lengths to estimate the aggregate reservation. For illustration, consider again the example in Figure 3.2, where flow 1 has a reservation rsv_1 of 5 Kbps. For the purpose of this example, we neglect the delay jitter, and assume that no packets are dropped inside the core. Suppose the inter-arrival times of the first four packets of flow 1 are 2 sec, 3 sec, and 4 sec, respectively. Since in this case the packet inter-arrival times at core routers are equal to the packet inter-departure times at the ingress, according to Eq. (3.3), the 2nd, 3rd, and 4th packet of flow 1 will carry in their headers $vl_1^2 = rsv_1 \times 2 = 10$ Kb, $vl_1^3 = rsv_1 \times 3 = 15$ Kb, and $vl_1^4 = rsv_1 \times 4 = 20$ Kb, respectively.

Next, note that the sum of the virtual lengths, $B_i(T)$, of all packets of flow i that arrive at a core router during an interval of length T, provides a fair approximation of *the amount of traffic that the flow is entitled to send during time T at its reserved rate*. Then, the reserved bandwidth of flow i, can be estimated as

$$\widehat{R_i} = \frac{B_i(T)}{T}. \qquad (3.5)$$

By extrapolation, a core router can estimate the aggregate reservation R on the outgoing link by simply computing $B(T)/T$, where $B(T)$ represents the sum of the virtual lengths of *all* packets that arrive during an interval of length T. Finally, it is worth noting that to perform this computation core routers do not need to maintain any per flow state – they just need to maintain a global variable, $B(T)$, that is updated every time a new packet arrives.

In the example shown in Figure 3.2, assume an averaging interval $T = 10$ sec (represented by the shaded area). Then we have $B(T) = B_1(T) + B_2(t) = 65$ Kb, which gives us an estimate of the aggregate reservation of

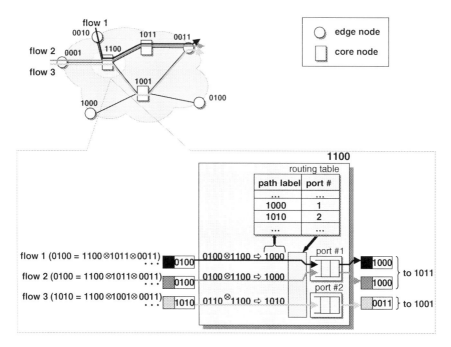

Fig. 3.3. Example illustrating the route pinning algorithm. Each packet contains in its header the path's label, defined as the xor over the identifiers of all routers on the remaining path to the egress. Upon packet arrival, the packet's header is updated to the label of the remaining path. The routing decisions are exclusively based on the packet's label (here the labels are assumed to be unique).

$\widehat{R} = B(T)/T = 6.5$ Kbps, which is "reasonably" close to the actual aggregate reservation $R = 7$ Kbps.

In Chapter 5 we derive an *upper bound* of the aggregate reservation along a link, instead of just an estimate. By using the upper bound we can guarantee that the link is never over-provisioned.

Example 3: Route Pinning Many applications such as traffic engineering and guaranteed services require that all packets of a flow to follow the same path. To achieve this goal, many solutions such as Tenet RCAP [8] and RSVP [128] require routers to maintain per flow state. In this section we present an alternate solution based on DPS in which core routers do not need to maintain any per flow state.

The key idea is to label a path by xor-ing the identifiers of all routers along the path. Consider a path $id_0, id_1, \ldots, id_n$, where id_j represents the identifier of the j-th router along the path. The label l of this path at router id_0 is then

$$l = id_1 \otimes id_2 \otimes \ldots \otimes id_n. \tag{3.6}$$

In the example in Figure 3.3, the label of flow 1 that enters the network at the ingress router 0010, and traverses routers 1100, 1011 and 0011 is simply $1100 \otimes 1011 \otimes 0011 = 0100$.

The DPS algorithm in this case is as follows: Each ingress router maintains a label l for every flow that traverses it. Upon packet arrival, ingress routers insert the label in the packet header.[1] Upon receiving a packet, a core router recomputes the label of the remaining path by xor-ing the label carried by the packet to its identifier. For example, when the first core router, identified by id_1, receives a packet with label l, it recomputes a new label as $l = l \otimes id_1$. Note that by doing so the new label represents exactly the identifier of the remaining path, i.e., $id_2 \otimes id_3 \otimes \ldots \otimes id_n$. Finally, the core router updates the label in the packet header, and uses the resulting label to forward the packet. Thus, a core router is not required to maintain per flow state, as it forwards each packet based on the label in its header.

Figure 3.3 gives an example of three flows that arrive to core router 1100, and exit the network through the same egress router 0011. However, while flows 1 and 2 are routed on identical paths, flow 3 is routed on a different path. When a packet arrives, the label in the packet header is updated by xor-ing it with the router identifier. Subsequently, the new label is used to route the packet. Note that only one entry is maintained for both flows 1 and 2.

3.2 Prototype Implementation

To demonstrate that it is possible to efficiently implement and deploy our solutions in today's IPv4 networks, we have developed a prototype implementation in FreeBSD v2.2.6. The prototype fully implements the guaranteed service as described in Chapter 5, arguably the most complex of all the solutions we describe in this dissertation. Without going into detail, we note that our solution to provide guaranteed services tries to closely approximate an idealized model in which each guaranteed flow traverses *dedicated* links of capacity r, where r is the flow reservation. Thus, in the idealized system, a flow with a reservation of 1 Mbps behaves as if it is the only flow in a network in which all links are of 1 Mbps.

The prototype runs in a test-bed consisting of 300 MHz and 400 MHz Pentium II PCs connected by point-to-point 100 Mbps Ethernets. The test-bed allows the configuration of a path with up to two core routers. Although we had complete control of our test-bed, and, due to resource constraints,

---

[1]Note that for simplicity, we do not present here how ingress routers obtain these labels. Also, we assume that path labels are unique, and therefore the routing decisions can be exclusively based on the path label. Finally, we do not discuss the impact of our scheme on address aggregation. We remove all these limitations in Chapter 6.

the scale of our experiments was rather small (e.g., the largest experiment involved just 100 flows), we have devoted special attention to making our implementation as general as possible. For example, while in the current implementation we re-use protocol space in the IP header to store the DPS state, we make sure that the modified fields can be fully restored by the egress router. In this way, the changes operated by the ingress and core routers on the packet header are completely transparent to the outside world. Similarly, while the limited scale of our experiments would have allowed us to use simple data structures to implement our algorithms, we go to great length to make sure that our implementation is scalable. For example, instead of using a simple linked list to implement the packet scheduler, we use a calendar queue together with a two-level priority queue to efficiently handle a very large number of flows (see Section 8.1).

For debugging and management purposes, we implemented full support for packet level monitoring. This allows us to visualize simultaneously and in real-time the throughputs and the delays experienced by flows at different points in the network. A key challenge when implementing such a fine grained monitoring functionality is to minimize the interferences with the system operations. We use two techniques to address this challenge. First, we off-load as much as possible of the processing of log data on an external machine. Second, we use raw IP to send *directly* the log data from router's kernel to the external machine. This way, we avoid context-switching between the kernel and the user level.

To easily configure our system, we have implemented a command line configuration tool. This tool allows us (1) to configure routers as ingress, egress, or core, (2) set-up, modify, and tear-down a reservation, and (3) set-up the monitoring parameters. To minimize the interferences between the configuration operations and data processing, we implement our tool on top of the Internet Control Management Protocol (ICMP). Again, by using ICMP, we avoid context-switching when configuring a router.

3.2.1 An Example

To illustrate how our entire package is working, in this section we present a simple example. We consider three flows traversing a three hop path in our test-bed (see Figure 3.4). The first router on the path (i.e., `aruba.cmcl.cs.cmu.edu`) is configured as an ingress router, while the next router (i.e., `cozumel.cmcl.cs.cmu.edu`) is configured as a core router. The link between the two routers is configured to 10 Mbps. The traffic of each flow is generated by a different end-host to eliminate the potential interferences. All flows are UDP, and send 1000 byte data packets. Flows 1 and 2 are guaranteed, while flow 3 is best-effort. More precisely,

– Flow 1 is a constant-bit rate (CBR) flow with an arrival rate of 1 Mbps, and a reservation of 1 Mbps.

– Flow 2 is ON-OFF, with the ON and OFF periods of 10 sec each. During the ON period the flow sends at 3 Mbps, while during the OFF period the flow does not send anything. The flow has a reservation of 3 Mbps.
– Flow 3 is CBR with an arrival rate of approximately 8 Mbps. Unlike flows 1 and 2, this flow is best-effort, i.e., it does not have any reservation.

Note that when all flows are active, the total offered load is about 12 Mbps, which exceeds the link capacity by 2 Mbps. As a result, during these time periods the ingress router is heavily congested.

To observe the behavior of our implementation during this experiment, we use an external machine (i.e., an IBM ThinkPad 560E notebook) to monitor the three flows at the end-points of the congested link: aruba and cozumel. Figure 3.5 shows a screen snapshot of our monitoring tool that plots the arrival rates and the delays experienced by the three flows at aruba, and cozumel, respectively, over a 56 sec time interval. The top-left plot shows the arrival rates of the three flows at aruba, while the top-right plot shows their arrival rates at cozumel. All rates represent averages over a 200 ms time period. As expected, flow 1, which has a reservation of 1 Mbps, and sends traffic at 1 Mbps, gets all its traffic through the congested link. This is illustrated by the straight line at 1 Mbps that appears in both plots. The same is true for flow 2; whenever it sends at 3 Mbps it gets its reservation. That is why the arrival rate of flow 2 looks identical in the two plots. In contrast, as shown in the top-right plot, flow 3 gets its service only when the link is uncongested, i.e., when flow 2 does not send anything. This is because flow 3 is best-effort, and therefore when both flows 1 and 2 fully use their reservations, flow 3 gets only the remaining bandwidth, which in this case is about 6 Mbps.

The bottom-left and the bottom-right plots in Figure 3.5 show the delays experienced by each flow at aruba, and cozumel, respectively. Each data point represents the maximum delay among all packets of a flow over a 200 ms time period. Note the different scales on the y-axis of the two plots. Next, we explain in more detail the results shown by these two plots.

Consider a flow i with reservation r_i that traverses a link of capacity C. Assume that the arrival rate of the flow never exceeds its reservation r_i, and that all packets have length l. Then, it can be shown that the worst case delay of a packet at a router is[2]

$$\frac{l}{r_i} + \frac{l}{C}, \tag{3.7}$$

Intuitively, the first term, l/r_i, represents how much it takes to transmit the packet in the ideal model in which the flow traverses dedicated links of capacity equal to its reservation r_i. The second term, l/C, represents the fact

---

[2]This result follows from Appendix B.2, which shows that the worst case delay of our packet scheduler, called Core Jitter Virtual Clock (CJVC) is identical to the worst case delay of Weighted Fair Queueing (WFQ) [79].

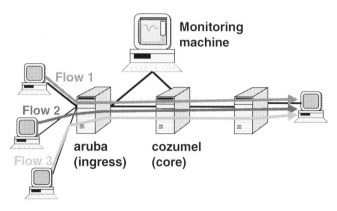

Fig. 3.4. The topology used in the experiment reported in Section 3.2. Flow 1 is CBR, has an arrival rate of 1 Mbps, and a reservation of 1 Mbps. Flow 2 is ON-OFF; it sends 3 Mbps during ON periods and doesn't send anything during OFF periods. The flow has a reservation of 3 Mbps. Flow 3 is best-effort and has an arrival rate of 8 Mbps. The link between `aruba` and `cozumel` is configured to 10 Mbps.

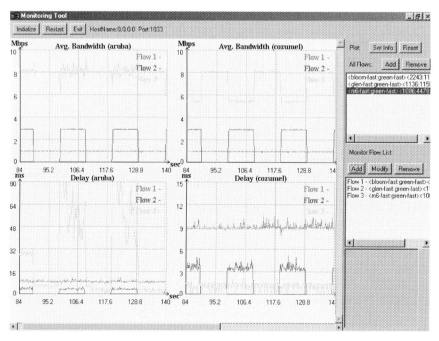

Fig. 3.5. A screen-shot of our monitoring tool that displays the real-time measurement results for the experiment shown in Figure 3.4. The top two plots show the arrival rate of each flow at `aruba` and `cozumel`; the bottom two plots show the delay experienced by each flow at the two routers.

that in a real system all flows share the link of capacity C, and that the packet transmission is not preemptive.

Since in our case $l \simeq 8400$ bits (this includes the packet headers), $C \simeq 10$ Mbps, $r_1 = 1$ Mbps, and $r_2 = 3$ Mbps, respectively, according to Eq. (3.7), the worst case delay of flow 1 is about 9.2 ms, and the worst case delay of flow 2 is about 3.6 ms. This is confirmed by the bottom two plots. As it can be seen, especially in the bottom-right plot, the measured delays for both flows are close to the theoretical values. The reason for which the measured values are *consistently* close to the worst case bounds is due to the non-work conserving nature of CJVC; even if the output link is idle, a packet of flow i can wait for up to l/r_i time in the rate-regulator before becoming eligible for transmission (see Section 5.3). The fact that the measured delays occasionally exceed the theoretical bounds is because FreeBSD is not a real-time operating systems. As a result, packet processing may take occasionally longer because unexpected interrupts, or system calls.

Finally, it is worth noting that when flow 2 is active, flow 3 experiences very large delays at the ingress router, i.e., over 80 ms. This is because during these time periods flow 3 is restricted to 6 Mbps, while its arrival rate is about 8 Mbps. In contrast, at the subsequent router, the packet delay of flow 3 is much smaller, i.e., under 2 ms. This is because the core router is no longer congested after the ingress has shed the extra traffic of flow 3. The reason the delay experienced by flow 3 is even lower than the delays experienced by the guaranteed flows is because, unlike these flows, flow 3 is not regulated, and therefore its packets are eligible for transmission as soon as they arrive.

3.3 Comparison to Intserv and Diffserv

To enhance the best effort service in the Internet, over the past decade the Internet Engineering Task Force (IETF) has proposed two major service architectures: Integrated Services (Intserv) [82] and Differentiated Services (Diffserv) [32]. In this section, we compare our SCORE architecture to both Intserv and Diffserv.

3.3.1 Intserv

As discussed in Section 2.3.3, Intserv is able to provide powerful and flexible services, such as Guaranteed [93] and Controlled-Load services [121], on a per flow basis. However this comes at the expense of a substantial increase in the complexity as compared to today's best-effort architecture. In particular, traditional Intserv solutions require routers to perform per flow admission control and maintain per flow state on the control path, and to perform per flow classification, scheduling, and buffer management on the data path. This complexity is arguably the main technical reason behind the failure to deploy Intserv in the Internet.

SCORE Advantages Most of the advantages of SCORE over Intserv derive from the fact that in a SCORE network, core routers do not need to maintain any per flow state. These advantages are:

– **Scalability** The fact that in a SCORE network routers do not need to maintain per flow state, significantly simplifies both the control and the data paths.

  On the data path, routers are no longer required to perform per flow classification, which is arguably the most complex operation on the data path (see Section 2.2.4). In addition, as we will show in Chapter 5, the complexity of buffer management and packet scheduling are greatly reduced.

  On the control path, as we have briefly discussed in Section 3.1.3, and as we will show in more detail in Chapter 5, by using DPS it is also possible to perform per flow admission control in a SCORE network. Ultimately, the absence of per flow state at core routers trivially eliminates one of the biggest challenges faced by stateful solutions in general, and Intserv in particular: maintaining the consistency of per flow state.

  In summary, the fact that core routers are not required to perform any per flow management, makes the SCORE architecture highly scalable with respect to the number of flows that traverse a router.

– **Robustness** Eliminating the need to maintain per flow state at core routers has another desirable consequence: the SCORE architecture is more *robust* in the presence of link and router failures.[3] This is due to the inherent difficulty of maintaining the consistency of dynamic, and replicated state in a distributed environment. As pointed out by Clark [22]: *"because of the distributed nature of the replication, algorithms to ensure robust replication are themselves difficult to build, and few networks with distributed state information provide any sort of protection against failure."* While soft-state mechanisms such as RSVP can alleviate this problem, there is a fundamental trade-off between message complexity and the time period during which the system is "allowed" to be in an inconsistent state: the shorter this period is, the greater the signalling overhead is.

Intserv Advantages While in this dissertation we show that SCORE can implement the strongest semantic service proposed by Intserv so far, i.e., the guaranteed service, it is still unclear whether SCORE can implement *all* possible per flow services that can be implemented by Intserv. To offer intuition as to what might be difficult to implement in a SCORE network, consider a service in which a flow is allocated a different share of the link capacity at each router along its path. In such a service a flow will receive on each link bandwidth in proportion to its share. To implement this service in SCORE, packets would have to carry complete information about flow shares at all

---

[3]In the case of a router, here we consider only *fail-stop* type of failures, i.e., the fact that the router (process) has failed is detectable by other routers (processes).

routers. Unfortunately, this may compromise the DPS algorithms' scalability as the state will increase with the path length. In Chapter 9 we discuss in more detail the potential limitations of SCORE with respect to per flow solutions.

In addition to the potential benefit of being able to implement more sophisticated per flow services, Intserv has two other advantages over SCORE:

- **Robustness** While the SCORE architecture is more robust in the case of fail-stop failures, Intserv is more robust in the case of partial reservation failures. To illustrate this point, consider a router misbehavior that inserts erroneous state in the packet headers. Since core routers process packets based on this state, such a failure can, at the limit, compromise the service in an entire SCORE domain. As an example, if in the network shown in Figure 3.3, router 1100 misbehaves by writing arbitrary information in the packet headers, this will affect not only the traffic that traverses router 1100, but also the traffic that traverses router 1001! This is due to the incorrect state carried by the packets of flow 3 that may ultimately affect the processing of packets of other flows that traverse router 1100. In contrast, with per flow solutions such a failure is strictly confined to the traffic that traverses the faulty router.

  However, in Chapter 7 we propose an approach called "verify-and-protect" that addresses this problem. The idea is to have routers statistically verify that the incoming packets carry consistent state. This enables routers to discover and isolate misbehaving end-hosts and routers.
- **Incremental Deployability** Since all routers in a domain have to implement the same algorithms, SCORE can be deployed only on a domain by domain basis. In contrast, Intserv solutions can be deployed on a router by router basis. However, it should be noted that for end-to-end services, this distinction is less important, as in the latter case (at least) all congested routers along the path have to deploy the service.

3.3.2 Diffserv

While at the architectural level both Diffserv and SCORE are similar in that they both try to push complexity out of the network core, they differ in two important aspects.

First, the approach advocated by the two architectures to implement new network services is different. The SCORE/DPS approach is *top-down*. We start with a service and then derive the algorithms that have to be implemented by a SCORE network in order to achieve the desired service. In contrast, Diffserv proposes a *bottom-up* approach. Diffserv standardizes a small number of per hop behaviors (such as priority service among a very small number of classes) to be implemented by router vendors. It is then the responsibility of the Internet Service Providers (ISPs) to configure their routers in order to achieve the desired service. Unfortunately, configuring these routers is a daunting task. At this point we are aware of no general framework that allows us to build sophisticated services such as providing flow protection by simply configuring a network of Diffserv routers.

Second, while in Diffserv, packet headers carry only limited information to differentiate among a small number of classes, in SCORE, packets carry fine grained per flow state which allows a SCORE network to implement far more sophisticated services.

Next we discuss the advantages and disadvantages of SCORE as compared to Diffserv.

SCORE Advantages The advantages of SCORE over Diffserv derive from the fact that the DPS algorithms operate at a much finer granularity both in terms of time and traffic aggregates: the state embedded in a packet can be highly dynamic, as it encodes the current state of the flow, rather than the static and global properties such as dropping or scheduling priority.

- **Service Granularity** While SCORE, like Intserv, can provide services on a per flow basis, Diffserv provides a coarser level of service differentiation among a small number of traffic classes. As a result, Diffserv cannot provide some useful services such as per flow bandwidth and delay guarantees or per flow protection.
- **Robustness** The extra state carried by the packets in SCORE can help to identify and isolate malfunctions in the network. In particular, with SCORE it is possible to detect a router that misbehaves by inserting erroneous state in the packet headers. To achieve this, in Chapter 7 we propose an approach, called "verify-and-protect", in which routers statistically verify whether the incoming packets are correctly marked. For example, in the case of CSFQ, a router can monitor a flow, estimate its rate, and then check this rate against the rate carried by the packet headers. If the two rates fall outside a "tolerable" range, this is an indication that an up-stream router misbehaves. Thus, the problem is confined to the routers on the path from the ingress where the flow enters the network up to the current router.

  In contrast, with Diffserv it is not possible to infer such an information. If, for example, a core router starts to drop a high percentage of premium packets this can be attributed to *any* router along any path from the ingress routers to the current router.

Diffserv Advantages

- **Data Path Processing Overhead** In Diffserv core routers process packets based on a small number of traffic classes. Upon packet arrival, a router classifies the packet, and then performs per class buffer management and scheduling. Since usually the number of classes is no larger than 10, packet processing can be very efficiently implemented. In contrast, in SCORE, packet processing can be more complex. For example, in the case of providing guaranteed services, each packet has an associated deadline, and the packets are served in the increasing order of their deadlines. However, as we will show in Chapter 5, the number of packets that have to be considered at one time is still much smaller than the number of flows. In particular, when the number of flows is larger than one million, the number of packets is at least two orders of magnitude smaller than the number of flows. We believe that this reduction is enough to allow packet processing at the line speed. Moreover, our other two solutions to provide per flow services, i.e., flow protection, and service differentiation of traffic aggregates over a large number of destinations, are no more complex than today's Diffserv solutions.

3.4 Summary

In this section we have described the main components of our solution to provide per flow services in a SCORE network architecture. To illustrate the key technique of our solution, i.e., Dynamic Packet State (DPS), we have presented the implementation of three per flow mechanisms in a SCORE network that (1) approximate Fair Queueing scheduling discipline, (2) provide per flow admission control, and (3) perform route pinning. In addition, we have compared our solution to two network architectures proposed by IETF to enhance the best-effort service (Intserv and Diffserv), and conclude that our solution achieves the best of the two worlds. In particular, it can provide services as powerful and as flexible as the ones implemented by Intserv, while having similar complexity and scalability as Diffserv.

4 Providing Flow Protection in SCORE

In this chapter we present the first illustration of our general solution described in Chapter 3. The goal is to provide *flow protection*, which is one of the most desirable enhancements of today's best-effort service. If deployed, flow protection would allow diverse end-to-end congestion control schemes to seamlessly coexist in the Internet, and protect well behaved traffic against malicious or ill-behaved traffic. The solution of choice to achieve flow protection is to have routers implement fair bandwidth allocation. Unfortunately, previous known implementations of fair bandwidth allocation require stateful networks, that is, require routers to maintain per flow state and perform per flow management. In this chapter, we present a solution to address this problem. In particular we use the Dynamic Packet State (DPS) technique to provide fair bandwidth allocations in a SCORE network. To the best of our knowledge this is the first solution to provide fair bandwidth allocation in a stateless core network.

The rest of this chapter is organized as follows. The next section presents the motivations behind the flow isolation and describes the fair bandwidth allocation approach to implement this service. Section 4.2 outlines our solution developed in the SCORE/DPS framework, called Core-Stateless Fair Queueing (CSFQ). Section 4.3 focusses on the details of CSFQ and its performance both absolute and relative, while Section 4.4 presents simulation results and compare CSFQ to several other schemes. Finally, Section 4.5 presents related work, and Section 4.6 concludes the chapter by summarizing our results.

4.1 Background

Because of their reliance on statistical multiplexing, data networks such as the Internet, must provide some mechanism to control network congestion. Network congestion occurs when the rate of the traffic arriving at a link exceeds the link capacity. The current Internet, which primarily uses FIFO queueing and drop-tail mechanisms in its routers, relies on end-to-end congestion control in which end-hosts reduce their transmission rates when they detect that the network is congested. The most widely utilized form of end-to-end congestion control is Transport Control Protocol (TCP) [57], which has been tremendously successful in preventing congestion collapse.

I. Stoica: Stateless Core, LNCS 2979, pp. 53-75, 2004.
© Springer-Verlag Berlin Heidelberg 2004

However, the effectiveness of this approach depends on one fundamental assumption: end-hosts *cooperate* by implementing *homogeneous* congestion control algorithms. In other words these algorithms produce similar bandwidth allocations if used in similar circumstances. In today's Internet this is equivalent to flows being "TCP-friendly", which means that "their arrival rate does not exceed that of any TCP connection in the same circumstances" [36].

While this was a reasonable assumption in the past when the Internet was primarily used by the research community, and the vast majority of traffic was TCP based, it is no longer true today. In particular, this assumption can be violated in three general ways. First, some applications are *unresponsive* in that they don't implement any congestion control algorithms at all. Most of the early multimedia and multicast applications, like *vat* [59], *nv* [40], *vic* [70], *wb* [58] and RealAudio fall into this category. Another example would be malicious users mounting denial of service attacks by blasting unresponsive traffic into the network. Second, some applications use congestion control algorithms that, while responsive, are not TCP-friendly. An example of such an algorithm is Receiver-driven Layered Multicast (RLM) [69].[1] Third, some users will cheat and use a non-TCP congestion control algorithm to get more bandwidth. An example of this would be using a modified form of TCP with, for instance, a larger initial window and window opening constants.

Starting with Nagle [74], many researchers observed that these problems can be overcome when routers have mechanisms that allocate bandwidth in a fair manner. Fair bandwidth allocation protects well-behaved flows from the ill-behaved (unfriendly) flows, and allows a diverse set of end-to-end congestion control policies to co-exist in the network [31]. To differentiate it from other approaches (see Section 4.5 for an alternative approach) that deal with the unfriendly flow problem we call this approach the *allocation* approach. It is important to note that the allocation approach does not demand that all flows adopt some universally standard end-to-end congestion control algorithm; flows can choose to respond to the congestion in whatever manner best suits them without harming other flows. Assuming that flows prefer not to have significant levels of packet drop, these allocation approaches give an incentive for flows to use end-to-end congestion control, because being unresponsive hurts their own performance.

While the allocation approach has many desirable properties for congestion control, it has yet to be deployed in the Internet. One of the main reasons behind this state of affairs is the implementation complexity. Until now, fair allocations were typically achieved by using *per flow* queueing mechanisms – such as Fair Queueing [31, 79] and its many variants [10, 45, 94] – or *per flow* dropping mechanisms such as Flow Random Early Drop (FRED) [67]. These

---

[1]Although our data in Section 4.4 showed RLM receiving less than its fair share, when we change the simulation scenario so that the TCP flow starts after all the RLM flows, it then receives less than half of its fair share. This hysteresis in the RLM versus TCP behavior was first pointed out to us by Steve McCanne [69].

mechanisms are significantly more complex to implement than the traditional FIFO queueing with drop-tail, which is the most widely implemented and deployed mechanism in routers today. In particular, fair allocation mechanisms inherently require the router to maintain state and perform operations on a per flow basis. For each packet that arrives at the router, the routers needs to *classify* the packet into a flow, update per flow state variables, and perform certain operations based on the per flow state. The operations can be as simple as deciding whether to drop or queue the packet (e.g., FRED), or as complex as manipulation of priority queues (e.g., Fair Queueing). While a number of techniques have been proposed to reduce the complexity of the per packet operations [9, 94, 99], and commercial implementations are available in some intermediate class routers, it is still unclear whether these algorithms can be cost-effectively implemented in high-speed backbone routers because all these algorithms still require packet classification and per flow state management.

In this chapter, we address the complexity problem by describing a solution based on Dynamic Packet State to provide fair bandwidth allocation within a SCORE domain. We call this solution Core-Stateless Fair Queueing (CSFQ) since the core routers keep no per flow state, but instead use the state that is carried in the packet labels.

4.2 Solution Outline

Existing solutions to provide fair bandwidth allocation require routers to maintain per flow state [10, 31, 45, 79, 94]. In this chapter, we present the first solution to achieve fair bandwidth allocation in a network in which core routers maintain no per flow state. Our solution is based on the generic approach described in Section 3.1.2. This approach consists of two steps. In the

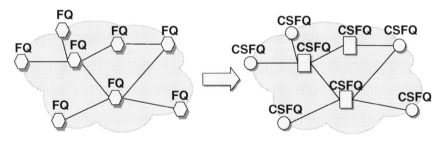

a) Reference Stateful Network   b) SCORE Network

Fig. 4.1. (a) A reference stateful network that provides fair bandwidth allocation; each node implements the Fair Queueing algorithm. (b) A SCORE network that approximates the service provided by the reference network; each node implements our algorithm, called *Core*-Stateless Fair Queueing (CSFQ)

first step we define a reference network that provides fair bandwidth allocation by having each node implement the Fair Queueing (see Figure 4.1(a))
algorithm. In the second step, we approximate the service provided by the
reference network within a SCORE network (see Figure 4.1(b)). To achieve
this we use the Dynamic Packet State (DPS) technique to implement a novel
algorithm, called *Core*-Stateless Fair Queueing (CSFQ), which approximates
the behavior of Fair Queueing.

With CSFQ, edge routers use per flow state to estimate the rate of each
incoming flow. Upon a packet arrival, the edge router classifies the packet to
the appropriate flow, updates the flow's rate estimate, and then labels the
packet with this estimate. In turn, core routers[2] implement FIFO queueing
with probabilistic dropping on input. The probability of dropping a packet
as it arrives at the queue is a function of the rate estimate carried in the
label and of the fair share rate at that router, which is estimated based on
measurements of the aggregate traffic. When the packet is forwarded the
router may update the estimate carried by the packet to reflect the eventual
change in the flow's rate due to packet dropping. In this way, CSFQ avoids
both the need to maintain per flow state and the need to use complicated
packet scheduling and buffering algorithms at core routers.

4.3 Core-Stateless Fair Queueing (CSFQ)

In this section we present our algorithm, called Core-Stateless Fair Queueing
(CSFQ), which approximates the behavior of Fair Queueing in a SCORE
network. To offer intuition about how CSFQ works, we first present the idealized bit-by-bit or *fluid* version of the probabilistic dropping scheme, and
then extend the algorithm to a practical packet-by-packet version.

4.3.1 Fluid Model Algorithm

We first consider a bufferless fluid model of a router with output link speed
C, where the flows are modelled as a continuous stream of bits. We assume
each flow's arrival rate $r_i(t)$ is known precisely. Max-min fair bandwidth
allocations are characterized by the fact that all flows that are bottlenecked
(i.e., have bits dropped) by this router have the same output rate. We call
this rate the *fair share rate* of the server; let $\alpha(t)$ be the fair share rate at
time t. In general, if max-min bandwidth allocations are achieved, each flow
i receives service at a rate given by $\min(r_i(t), \alpha(t))$. Let $A(t)$ denote the total
arrival rate: $A(t) = \sum_{i=1}^{n} r_i(t)$. If $A(t) > C$ then the fair share $\alpha(t)$ is the
unique solution to

$$C = \sum_{i=1}^{n} \min(r_i(t), \alpha(t)), \qquad (4.1)$$

---

[2]Note that Example 1 in Section 3.1.3 outlines the CSFQ algorithm as implemented by core routers.

If $A(t) \leq C$ then no bits are dropped and we will, by convention, set $\alpha(t) = \max_i r_i(t)$.

If $r_i(t) \leq \alpha(t)$, i.e., flow i sends no more than the server's fair share rate, all of its traffic will be forwarded. If $r_i(t) > \alpha(t)$, then a fraction $\frac{r_i(t)-\alpha(t)}{r_i(t)}$ of its bits will be dropped, so it will have an output rate of exactly $\alpha(t)$. This suggests a very simple probabilistic forwarding algorithm that achieves fair allocation of bandwidth: each incoming bit of flow i is dropped with the probability

$$\max\left(0, 1 - \frac{\alpha(t)}{r_i(t)}\right) \tag{4.2}$$

When these dropping probabilities are used, the arrival rate of flow i at the next hop is given by $\min[r_i(t), \alpha(t)]$.

4.3.2 Packet Algorithm

The above algorithm is defined for a bufferless fluid system in which the arrival rates are known exactly. Our task now is to extend this approach to the situation in real routers where transmission is packetized, there is substantial buffering, and the arrival rates are not known.

We still employ a drop-on-input scheme, except that now we drop packets rather than bits. Because the rate estimation (described below) incorporates the packet size, the dropping probability is independent of the packet size and depends only, as above, on the rate $r_i(t)$ and fair share rate $\alpha(t)$.

We are left with two remaining challenges: estimating the rates $r_i(t)$ and the fair share $\alpha(t)$. We address these two issues in turn in the next two subsections, and then discuss the rewriting of the labels. Pseudocode reflecting this algorithm is described in Figures 4.3 and 4.4.

Computation of Flow Arrival Rate Recall that in our architecture, the rates $r_i(t)$ are estimated at the edge routers and then these rates are inserted into the packet labels. At each edge router, we use exponential averaging to estimate the rate of a flow. Let t_i^k and l_i^k be the arrival time and length of the kth packet of flow i. The estimated rate of flow i, r_i, is updated every time a new packet is received:

$$r_i^{new} = (1 - e^{-T_i^k/K})\frac{l_i^k}{T_i^k} + e^{-T_i^k/K}r_i^{old}, \tag{4.3}$$

where $T_i^k = t_i^k - t_i^{k-1}$ and K is a constant. We discuss the rationale for using the form $e^{-T_i^k/K}$ for the exponential weight in Section 4.3.7.

Link Fair Rate Estimation In this section, we present an estimation algorithm for $\alpha(t)$. To give intuition, consider again the fluid model in Section 4.3.1 where the arrival rates are known exactly, and assume the system

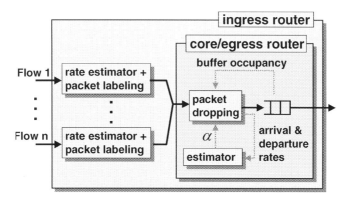

Fig. 4.2. The architecture of the output port of an edge router, and a core router, respectively.

performs the probabilistic dropping algorithm according to Eq. (4.2). Then, the rate with which the algorithm accepts packets is a function of the current estimate of the fair share rate, which we denote by $\widehat{\alpha}(t)$. Letting $F(\widehat{\alpha}(t))$ denote this acceptance rate, we have

$$F(\widehat{\alpha}(t)) = \sum_{i=1}^{n} \min\left(r_i(t), \widehat{\alpha}(t)\right). \qquad (4.4)$$

Note that $F(\cdot)$ is a continuous, nondecreasing, concave, and piecewise-linear function of $\widehat{\alpha}$. If the link is congested $(A(t) > C)$ we choose $\widehat{\alpha}(t)$ to be the unique solution to $F(x) = C$. If the link is not congested $(A(t) < C)$ we take $\widehat{\alpha}(t)$ to be the largest rate among the flows that traverse the link, i.e., $\widehat{\alpha}(t) = \max_{1 \leq i \leq n}(r_i(t))$. From Eq. (4.4) note that if we knew the arrival rates $r_i(t)$ we could then compute $\alpha(t)$ directly. To avoid having to keep such per flow state, we seek instead to implicitly compute $\widehat{\alpha}(t)$ by using only aggregate measurements of F and A.

We use the following heuristic algorithm with three aggregate state variables: $\widehat{\alpha}$, the estimate for the fair share rate; $\widehat{A}$, the estimated aggregate arrival rate; $\widehat{F}$, the estimated rate of the accepted traffic. The last two variables are updated upon the arrival of each packet. For $\widehat{A}$ we use exponential averaging with a parameter e^{-T/K_α} where T is the inter-arrival time between the current and the previous packet:

$$\widehat{A}_{new} = (1 - e^{-T/K_\alpha})\frac{l}{T} + e^{-T/K_\alpha}\widehat{A}_{old} \qquad (4.5)$$

where $\widehat{A}_{old}$ is the value of $\widehat{A}$ before the updating. We use an analogous formula to update $\widehat{F}$.

The updating rule for $\widehat{\alpha}$ depends on whether the link is congested or not. To filter out the estimation inaccuracies due to exponential smoothing we

```
on packet p arrival
  if (edge router)
    i =classify(p);
    p.label = estimate_rate(rᵢ, p); /* use Eq. (4.3) */
  prob =max(0, 1 − α/p.label);
  if (prob >unif_rand(0, 1))
    α =estimate_α (p, 1);
    drop(p);
  else
    α =estimate_α (p, 0);
    enqueue(p);
  if (prob > 0)
    p.label = α; /* relabel p */
```

Fig. 4.3. The pseudocode of CSFQ.

use a window of size K_c. A link is assumed to be *congested*, if $\widehat{A} \geq C$ at all times during an interval of length K_c. Conversely, a link is assumed to be *uncongested*, if $\widehat{A} \leq C$ at all times during an interval of length K_c. The value $\widehat{\alpha}$ is updated only at the end of an interval in which the link is either congested or uncongested according to these definitions. If the link is congested then $\widehat{\alpha}$ is updated based on the equation $F(\widehat{\alpha}) = C$. We approximate $F(\cdot)$ by a linear function that intersects the origin and has slope $\widehat{F}/\widehat{\alpha}_{old}$. This yields

$$\widehat{\alpha}_{new} = \widehat{\alpha}_{old} \frac{C}{\widehat{F}} \qquad (4.6)$$

If the link is not congested, $\widehat{\alpha}_{new}$ is set to the largest rate of any active flow (i.e., the largest label seen) during the last K_c time units. The value of $\widehat{\alpha}_{new}$ is then used to compute dropping probabilities, according to Eq. (4.2). For completeness, we give the pseudocode of the CSFQ algorithm in Figure 4.4.

We now describe two minor amendments to this algorithm related to how the buffers are managed. The goal of estimating the fair share $\widehat{\alpha}$ is to match the accepted rate to the link bandwidth. Due to estimation inaccuracies, load fluctuations between $\widehat{\alpha}$'s updates, and the probabilistic nature of our algorithm, the accepted rate may occasionally exceed the link capacity. While ideally the router's buffers can accommodate the extra packets, occasionally the router may be forced to drop the incoming packet due to lack of buffer space. Since drop-tail behavior will defeat the purpose of our algorithm, and may exhibit undesirable properties in the case of adaptive flows such as TCP [37], it is important to limit its effect. To do so, we use a simple heuristic: every time the buffer overflows, $\widehat{\alpha}$ is decreased by a small fixed percentage (taken to be 1% in our simulations). Moreover, to avoid overcorrection, we make sure that during consecutive updates $\widehat{\alpha}$ does not decrease by more than 25%.

In addition, since there is little reason to consider a link congested if the buffer is almost empty, we apply the following rule. If the link becomes uncongested by the test in Figure 4.4, then we assume that it remains uncongested as long as the buffer occupancy is less than some predefined threshold. In the current implementation we use a threshold that is half of the total buffer capacity.

estimate_α $(p, dropped)$
  estimate_rate$(\widehat{A}, p)$; /* est. arrival rate (use Eq. (4.5)) */
  if $(dropped == FALSE)$
    estimate_rate$(\widehat{F}, p)$; /* est. accepted traffic rate */
  if $(\widehat{A} \geq C)$
    if $(congested == FALSE)$
      $congested = TRUE$;
      $start\_time = crt\_time$;
    else
      if $(crt\_time > start\_time + K_c)$
      $\widehat{\alpha} = \widehat{\alpha} \times C/\widehat{F}$;
      $start\_time = crt\_time$;
  else /* $\widehat{A} < C$ */
    if $(congested == TRUE)$
      $congested = FALSE$;
      $start\_time = crt\_time$;
      $tmp\_\alpha = 0$; /* use to compute new α */
    else
      if $(crt\_time < start\_time + K_c)$
        $tmp\_\alpha =$**max**$(tmp\_\alpha, p.label)$;
      else
      $\widehat{\alpha} = tmp\_\alpha$;
      $start\_time = crt\_time$;
      $tmp\_\alpha = 0$;
  return $\widehat{\alpha}$;

Fig. 4.4. The pseudocode of CSFQ (fair rate estimation).

Label Rewriting Our rate estimation algorithm in Section 4.3.2 allows us to label packets with their flow's rate as they enter the SCORE domain. Our packet dropping algorithm described in Section 4.3.2 allows us to limit flows to their fair share of the bandwidth. After a flow experiences significant losses at a congested link inside the domain, however, the packet labels are no longer an accurate estimate of its rate. We cannot rerun our estimation algorithm, because it involves per flow state. Fortunately, as noted in Section 4.3.1, the outgoing rate is merely the incoming rate or the fair rate, α, whichever is smaller. Therefore, we rewrite the the packet label L as

$$L_{new} = \min(L_{old}, \alpha), \tag{4.7}$$

By doing so, the outgoing flow rates will be properly represented by the packet labels.

4.3.3 Weighted CSFQ

The CSFQ algorithm can be extended to support flows with different weights. Let w_i denote the weight of flow i. Returning to our fluid model, the meaning of these weights is that we say a *fair* allocation is one in which all bottlenecked flows have the same value for $\frac{r_i}{w_i}$. Then, if $A(t) > C$, the *normalized* fair rate $\alpha(t)$ is the unique value such that $\sum_{i=1}^{n} w_i \min\left(\alpha, \frac{r_i}{w_i}\right) = C$. The expression for the dropping probabilities in the weighted case is $\max\left(0, 1 - \alpha \frac{w_i}{r_i}\right)$. The only other major change is that the label is now r_i/w_i, instead simply r_i. Finally, without going into detail we note that the weighted packet-by-packet version is virtually identical to the corresponding version of the plain CSFQ algorithm.

It is also important to note that with weighted CSFQ we can only approximate a reference network in which each flow has the same weight at all routers along its path. That is, our algorithm cannot accommodate situations where the relative weights of flows differ from router to router within a domain. However, even with this limitation, weighted CSFQ may prove a valuable mechanism in implementing differential services, such as the one proposed in [116].

4.3.4 Performance Bounds

We now present the main theoretical result for CSFQ. For generality, this result is given for weighted CSFQ. The proof is given in Appendix A.

Our algorithm is built around several estimation procedures, and thus is inherently inexact. One natural concern is whether a flow can purposely "exploit" these inaccuracies to get more than its fair share of bandwidth. We cannot answer this question with full generality, but we can analyze a simplified situation where the normalized fair share rate, α, is held fixed and there is no buffering, so the drop probabilities are precisely given by Eq. (4.2). In addition, we assume that when a packet arrives, a fraction of that packet equal to the flow's forwarding probability is transmitted. Note that during any time interval $[t_1, t_2)$ a flow with weight w is entitled to receive at most $w\alpha(t_2 - t_1)$ service time; we call any amount above this the *excess service*. This excess service can be bound, independent of both the arrival process and the length of the time interval during which the flow is active. The bound does depend crucially on the maximal rate, R, at which a flow packets can arrive at a router (limited, for example, by the speed of the flow's access link); the smaller this rate R, the tighter the bound.

Theorem 1 *Consider a link with a constant normalized fair rate α, and a flow with weight w. Then, the excess service received by a flow with weight w that sends at a rate no larger than R is bounded above by*

$$r_\alpha K \left(1 + ln\frac{R}{r_\alpha} \right) + l_{\max}, \tag{4.8}$$

where $r_\alpha = \alpha w$, and l_{max} represents the maximum length of a packet.

By bounding the excess service, we have shown that in this idealized setting, the asymptotic throughput cannot exceed the fair share rate. Thus, flows can only exploit the system over short time scales; they are limited to their fair share over long time scales.

4.3.5 Implementation Complexity

At core routers, both the time and space complexity of our algorithm are constant with respect to the number of competing flows, and thus we think CSFQ could be implemented in very high speed core routers. At each edge router CSFQ needs to maintain per flow state. Upon the arrival of each packet, the edge router needs to (1) classify the packet to a flow, (2) update the fair share rate estimation for the corresponding outgoing link, (3) update the flow rate estimation, and (4) label the packet. All these operations with the exception of packet classification can be efficiently implemented today.

Efficient and general-purpose packet classification algorithms are still under active research. We expect to leverage these results. We also note that packet classification at ingress nodes is needed for a number of other purposes, such as in the context of Multiprotocol Label Switching (MPLS) [17] or for accounting purposes. Therefore, the classification required for CSFQ may not be an extra cost. In addition, edge routers typically not on the high-speed backbone links pose no problem as classification at moderate speeds is quite practical.

4.3.6 Architectural Considerations

We have used the term flow without defining what we mean. This was intentional, as the CSFQ approach can be applied to varying degrees of flow granularity; that is, what constitutes a flow is arbitrary as long as all packets in the flow follow the same path within the core. For convenience, flow as used here is implicitly defined as a source-destination pair, but one could easily assign fair rates to many other granularities such as source-destination-ports. Moreover, the unit of "flow" can vary from domain to domain as long as the rates are re-estimated when entering a new domain.

Similarly, we have not been precise about the size of the SCORE domains. In one extreme, we could take each router as a domain and estimate rates at every router; this would allow us to avoid the use of complicated per flow scheduling and dropping algorithms, but would require per flow classification. Another possibility is that ISPs could extend their SCORE domain to the very edge of their network, having their edge routers at the points where customer's packets enter the ISP's network. Building on the previous scenario, multiple ISPs could combine their domains so that classification and estimation did not have to be performed at ISP-ISP boundaries. The key obstacle here is one of trust between ISPs.

4.3.7 Miscellaneous Details

Having presented the basic CSFQ algorithm, we now return to discuss a few aspects in more detail.

We have used exponential averaging to estimate the arrival rate in Eq. (4.3). However, instead of using a constant exponential weight we used $e^{-T/K}$, where T is the inter-packet arrival time and K is a constant. Our motivation was that $e^{-T/K}$ more closely reflects a fluid averaging process which is independent of the packetizing structure. More specifically, it can be shown that if a constant weight is used, the estimated rate will be sensitive to the packet length distribution and there are pathological cases where the estimated rate differs from the real arrival rate by a factor; this would allow flows to exploit the estimation process and obtain more than their fair share. In contrast, by using a parameter of $e^{-T/K}$, the estimated rate will asymptotically converge to the real rate, and this allows us to bound the excess service that can be achieved (as in Theorem 1). We used a similar averaging process in Eq. (4.5) to estimate the total arrival rate A.

The choice of K in the above expression $e^{-T/K}$ presents us with several tradeoffs. First, while a smaller K increases system responsiveness to rapid rate fluctuations, a larger K better filters noise and avoids potential system instability. Second, K should be large enough such that the estimated rate, calculated at the edge of the network, remains reasonably accurate after a packet traverses multiple links. This is because the delay-jitter changes the packets' inter-arrival pattern, which may result in an increased discrepancy between the estimated rate (received in the packets' labels) and the real rate. To counteract this effect, as a rule of thumb, K should be one order of magnitude larger that the delay-jitter experienced by a flow over a time interval of the same size, K. Third, K should be no larger than the average duration of a flow. Based on this constraints, an appropriate value for K would be between 100 and 500 ms.

4.4 Simulation Results

In this section we evaluate our algorithm by simulation. To provide some context, we compare CSFQ's performance to three additional algorithms. Two of these, FIFO and RED, represent baseline cases where routers do not attempt to achieve fair bandwidth allocations. The other two algorithms, FRED and DRR, represent different approaches to achieving fairness.

- FIFO (First In First Out) - Packets are served in a first-in first-out order, and the buffers are managed using a simple drop-tail strategy; i.e., incoming packets are dropped when the buffer is full.
- RED (Random Early Detection) - Packets are served in a first-in first-out order, but the buffer management is significantly more sophisticated than drop-tail. RED [37] starts to probabilistically drop packets long before the buffer is full, providing early congestion indication to flows which can then gracefully back-off before the buffer overflows. RED maintains two buffer thresholds. When the exponentially averaged buffer occupancy is smaller than the first threshold, no packet is dropped, and when the exponentially averaged buffer occupancy is larger than the second threshold all packets are dropped. When the exponentially averaged buffer occupancy is between the two thresholds, the packet dropping probability increases linearly with buffer occupancy.
- FRED (Fair Random Early Drop) - This algorithm extends RED to provide some degree of fair bandwidth allocation [67]. To achieve fairness, FRED maintains state for all flows that have at least one packet in the buffer. Unlike RED where the dropping decision is based only on the buffer state, in FRED dropping decisions are based on this flow state. Specifically, FRED preferentially drops a packet of a flow that has either (1) had many packets dropped in the past, or (2) a queue larger than the average queue size. FRED has two variants (which we will call FRED-1 and FRED-2). The main difference between the two is that FRED-2 guarantees to each flow a minimum number of buffers. As a general rule, FRED-2 performs better than FRED-1 only when the number of flows is large. In the following data, when we don't distinguish between the two, we are quoting the results from the version of FRED which performed the best.
- DRR (Deficit Round Robin) - This algorithm represents an efficient implementation of the well-known weighted fair queueing (WFQ) discipline. The buffer management scheme assumes that when the buffer is full the packet from the longest queue is dropped. DRR is the only one of the four to use a sophisticated per-flow queueing algorithm, and thus achieves the highest degree of fairness.

These four algorithms represent four different levels of complexity. DRR and FRED have to classify incoming flows, whereas FIFO and RED do not. In addition, DRR has to implement its packet scheduling algorithm (whereas the rest all use first-in-first-out scheduling). CSFQ edge routers have complexity comparable to FRED, and CSFQ core routers have complexity comparable to RED.

We have examined the behavior of CSFQ under a variety of conditions. We use an assortment of traffic sources (mainly CBR and TCP, but also some on-off sources) and topologies. For space reasons, we only report on a small sampling of the simulations we have run; a fuller set of tests, and the scripts used to run them, is available at http://www.cs.cmu.edu/~istoica/csfq. All simulations were performed in ns-2 [78], which provides accurate packet-level implementation for various network protocols, such as TCP and RLM (Receiver-driven Layered Multicast) [69], and various buffer management and scheduling algorithms, such as RED and DRR.

Unless otherwise specified, we use the following parameters for the simulations in this section. Each output link has a latency of 1 ms, a buffer of 64 KB, and a buffer threshold for CSFQ of 16 KB. In the RED and FRED cases, the first threshold is set to 16 KB, while the second one is set to 32 KB. The averaging constant used in estimating the flow rate is $K = 100$ ms, while the averaging constant used in estimation the fair rate α is $K_\alpha = 200$ ms. Finally, in all topologies we use the first router (gateway) on the path of a flow is always assumed to be the edge router; all other routers are assumed without exception to be core routers.

We simulated the other four algorithms to give us benchmarks against which to assess these results. We use DRR as our model of fairness and use the baseline cases, FIFO and RED, as representing the (unfair) status quo. The goal of these experiments is determine where CSFQ sits between these two extremes. FRED is a more ambiguous benchmark, being somewhat more complex than CSFQ, but not as complex as DRR.

In general, we find that CSFQ achieves a reasonable degree of fairness, significantly closer to DRR than to FIFO or RED. CSFQ's performance is typically comparable to FRED's, although there are a few situations where CSFQ significantly outperforms FRED. There are a large number of experiments and each experiment involves rather complex dynamics. Due to space limitations, in the sections that follow we will merely highlight a few important points and omit detailed explanations of the dynamics.

4.4.1 A Single Congested Link

We first consider a single congested link shared by N flows (see Figure 4.5(a)). We performed three related experiments.

In the first experiment, we have 32 CBR flows, indexed from 0, where flow i sends $i + 1$ times more than its fair share of 0.3125 Mbps. Thus flow 0 sends 0.3125 Mbps, flow 1 sends 0.625 Mbps, and so on. Figure 4.5(b) shows the average throughput of each flow over a 10 sec interval; FIFO, RED, and FRED-1 fail to ensure fairness, with each flow getting a share proportional to its incoming rate, while DRR is extremely effective in achieving a fair bandwidth distribution. CSFQ and FRED-2 achieve a less precise degree of fairness; for CSFQ the throughputs of all flows are between -11% and $+12\%$ of the ideal value.

In the second experiment we consider the impact of an ill-behaved CBR flow on a set of TCP flows. More precisely, the traffic of flow 0 comes from

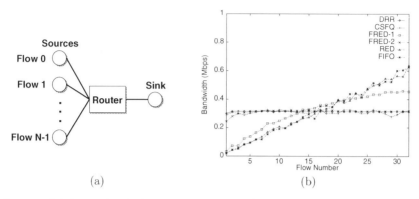

Sources

Flow 0

Flow 1   Sink

        Router

Flow N-1

(a)   (b)

Fig. 4.5. (a) A 10 Mbps link shared by N flows. (b) The average throughput over 10 sec when N = 32, and all flows are CBRs. The arrival rate for flow i is (i + 1) times larger than its fair share. The flows are indexed from 0.

a CBR source that sends at 10 Mbps, while all the other flows (from 1 to 31) are TCPs. Figure 4.6(a) shows the throughput of each flow averaged over a 10 sec interval. The only two algorithms that can most effectively contain the CBR flow are DRR and CSFQ. Under FRED the CBR flow gets almost 1.8 Mbps – close to six times more than its fair share – while the CBR only gets 0.396 Mbps and 0.355 Mbps under DRR and CSFQ, respectively. As expected, FIFO and RED perform poorly, with the CBR flow getting over 8 Mbps in both cases.

In the final experiment, we measure how well the algorithms can protect a single TCP flow against multiple ill-behaved flows. We perform 31 simulations, each for a different value of N, $N = 1 \ldots 31$. In each simulation we take one TCP flow and N CBR flows; each CBR sends at twice its fair share rate of $\frac{10}{N+1}$ Mbps. Figure 4.6(b) plots the ratio between the average throughput of the TCP flow over 10 sec and the total bandwidth it should receive as a function of the total number of flows in the system $N + 1$. There are three points of interest. First, DRR performs very well when there are less than 22 flows, but its performances decreases afterwards because then the TCP flow's buffer share is less than three buffers which is known to significantly affect its throughput. Second, CSFQ performs better than DRR when the number of flows is large. This is because CSFQ is able to cope better with the TCP burstiness by allowing the TCP flow to have more than two packets buffered for short time intervals. Finally, across the entire range, CSFQ provides similar or better performance as compared to FRED.

4.4.2 Multiple Congested Links

We now analyze how the throughput of a well-behaved flow is affected when the flow traverses more than one congested link. We performed two experi-

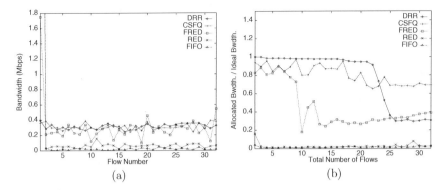

Fig. 4.6. (a) The throughputs of one CBR flow (0 indexed) sending at 10 Mbps, and of 31 TCP flows sharing a 10 Mbps link. (b) The normalized bandwidth of a TCP flow that competes with N CBR flows sending at twice their allocated rates, as a function of N.

ments based on the topology shown in Figure 4.7. All CBRs, except CBR-0, send at 2 Mbps. Since each link in the system has 10 Mbps capacity, this will result in all links between routers being congested.

In the first experiment, we have a CBR flow (denoted CBR-0) sending at its fair share rate of 0.909 Mbps. Figure 4.8(a) shows the fraction of CBR-0's traffic that is forwarded, versus the number of congested links. CSFQ and FRED perform reasonably well, although not quite as well as DRR.

In the second experiment we replace CBR-0 with a TCP flow. Similarly, Figure 4.8(b) plots the normalized TCP throughput against the number of congested links. Again, DRR and CSFQ prove to be effective. In comparison, FRED performs significantly worse though still much better than RED and FIFO. The reason is that while DRR and CSFQ tries to allocate bandwidth fairly among competing flows during congestion, FRED tries to allocate the buffer fairly. Flows with different end-to-end congestion control algorithms will achieve different throughputs even if routers try to fairly allocate the buffer. In the case of Figure 4.8(a), all sources are CBR, i.e., none are adopting any end-to-end congestion control algorithms, FRED provides performance similar to CSFQ and DRR. In the case of Figure 4.8(b), a TCP flow is competing with multiple CBR flows. Since the TCP flow slows down during congestion while CBQ does not, it achieves significantly less throughput than a competing CBR flow.

4.4.3 Coexistence of Different Adaptation Schemes

In this experiment we investigate the extent to which CSFQ can deal with flows that employ different adaptation schemes. Receiver-driven Layered Multicast (RLM) [69] is an adaptive scheme in which the source sends the information encoded into a number of layers (each to its own multicast group)

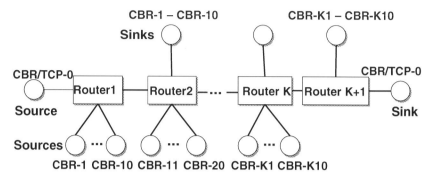

Fig. 4.7. Topology for analyzing the effects of multiple congested links on the throughput of a flow. Each link has ten cross flows (all CBRs). All links have 10 Mbps capacities. The sending rates of all CBRs, excepting CBR-0, are 2 Mbps, which leads to all links between routers being congested.

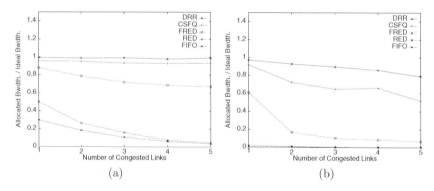

Fig. 4.8. (a) The normalized throughput of CBR-0 as a function of the number of congested links. (b) The same plot when CBR-0 is replaced by a TCP flow.

and the receiver joins or leaves the groups associated with the layers based on how many packet drops it is experiencing. We consider a 4 Mbps link traversed by one TCP and three RLM flows. Each source uses a seven layer encoding, where layer i sends 2^{i+4} Kbps; each layer is modeled by a CBR traffic source. The fair share of each flow is 1Mbps. In the RLM case this will correspond to each receiver subscribing to the first five layers. [3]

The average receiving rates averaged over 1 sec intervals for each algorithm are plotted in Figure 4.9. We have conducted two separate simulations of CSFQ.[4] In the first one, we have used the same averaging constant as in the rest of this chapter: $K = 100$ ms, and $K_\alpha = 200$ ms. Here, one RLM flow does not get its fair share (it is one layer below where it should be). We

---

[3]More precisely, we have $\sum_{i=1}^{5} 2^{i+4}$ Kbps $= 0.992$ Mbps.
[4]See also [69] for additional simulations of RLM and TCP.

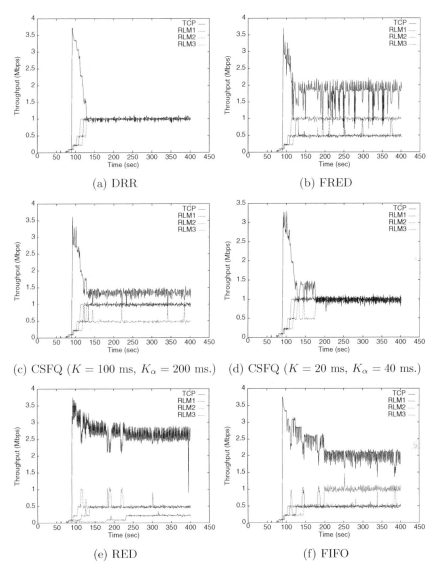

Fig. 4.9. The throughputs of three RLM flows and one TCP flow on a 4 Mbps link.

think this is due to the bursty behavior of the TCP that is not detected by
CSFQ soon enough, allowing the TCP to opportunistically grab more bandwidth than its share at the expense of less aggressive RLM flows. To test
this hypothesis, we have changed the averaging time intervals to $K = 20$ ms,
and $K_\alpha = 40$ ms, respectively, which result in TCP flow bandwidth being
restricted much earlier. As shown in Figure 4.9(d), with these parameters all
flows receive roughly 1 Mbps.

An interesting point to notice is that FRED does not provide fair bandwidth allocation in this scenario. Again, as discussed in Section 4.4.2, this
is due to the fact that RLM and TCP use different end-to-end congestion
control algorithms.

Finally, we note that we have performed two other similar experiments
(not included here due to space limitations): one in which the TCP flow is
replaced by a CBR that sends at 4 Mbps, and another in which we have
both the TCP and the CBR flows together along with the three RLM flows.
The overall results were similar, except that in both experiments *all* flows
received their shares under CSFQ when using the original settings for the
averaging intervals, i.e., $K = 100$ ms and $K_\alpha = 200$ ms. In addition, in some
of these other experiments where the RLM flows are started before the TCP,
the RLM flows get more than their share of bandwidth when RED and FIFO
are used.

4.4.4 Different Traffic Models

So far we have only considered CBR and TCP traffic sources. We now look
at two additional source models with greater degrees of burstiness. We again
consider a single 10 Mbps congested link. In the first experiment, this link
is shared by one ON-OFF source and 19 CBRs that send at exactly their
share, 0.5 Mbps. The ON and OFF periods of the ON-OFF source are both
drawn from exponential distributions with means of 200 ms and 19*200 ms
respectively. During the ON period the ON-OFF source sends at 10 Mbps.
Note that the ON-time is on the same order as the averaging interval $K =
200ms$ for CSFQ's rate estimation algorithm, so this experiment is designed
to test to what extent CSFQ can react over short time scales.

Algorithm	delivered	dropped
DRR	1080	3819
CSFQ	1000	3889
FRED	1064	3825
RED	2819	2080
FIFO	3771	1128

Table 4.1. Statistics for an ON-OFF flow with 19 Competing CBRs flows (all
numbers are in packets)

The ON-OFF source sent 4899 packets over the course of the experiment. Table 4.1 shows the number of packets from the ON-OFF source dropped at the congested link. The DRR results show what happens when the ON-OFF source is restricted to its fair share at all times. FRED and CSFQ also are able to achieve a high degree of fairness.

Our next experiment simulates Web traffic. There are 60 TCP transfers whose inter-arrival times are exponentially distributed with the mean of 0.1 ms, and the length of each transfer is drawn from a Pareto distribution with a mean of 40 packets (1 packet = 1 KB) and a shaping parameter of 1.06. These values are consistent with those presented by Crovella and Bestavros [27]. In addition, there is a single 10 Mbps CBR flow.

Algorithm	mean time	std. dev
DRR	46.38	197.35
CSFQ	88.21	230.29
FRED	73.48	272.25
RED	790.28	1651.38
FIFO	1736.93	1826.74

Table 4.2. The mean transfer times (in ms) and the corresponding standard deviations for 60 short TCPs in the presence of a CBR flow that sends at the link capacity, i.e., 10 Mbps.

Table 4.2 presents the mean transfer time and the corresponding standard deviations. Here, CSFQ and FRED do less well than DRR, but one order of magnitude better than FIFO and RED.

4.4.5 Large Latency

Algorithm	mean	std. dev
DRR	5857.89	192.86
CSFQ	5135.05	175.76
FRED	4967.05	261.23
RED	628.10	80.46
FIFO	379.42	68.72

Table 4.3. The mean throughput (in packets) and standard deviation for 19 TCPs in the presence of a CBR flow along a link with propagation delay of 100 ms. The CBR sends at the link capacity of 10 Mbps.

All of our experiments so far have had minimal latencies. In this experiment we again consider a single 10 Mbps congested link, but now the flows have propagation delays of 100 ms in getting to the congested link. The load is comprised of one CBR that sends at the link capacity and 19 TCP flows. Table 4.3 shows the mean number of packets forwarded for each TCP flow during a 100 sec time interval. CSFQ and FRED both perform reasonably well.

4.4.6 Packet Relabeling

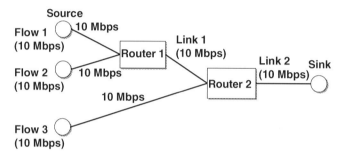

Fig. 4.10. Simulation scenario for the packet relabeling experiment. Each link has 10 Mbps capacity, and a propagation delay of 1 ms.

Traffic	Flow 1	Flow 2	Flow 3
CBR	3.267	3.262	3.458
TCP	3.232	3.336	3.358

Table 4.4. The throughputs resulting from CSFQ averaged over 10 sec for the three flows in Figure 4.10 along link 2.

Recall that when the dropping probability P of a packet is non-zero we relabel it with a new label where $L_{new} = (1 - P)L_{old}$ so that the label of the packet will reflect the new rate of the flow. To test how well this works in practice, we consider the topology in Figure 4.10, where each link is 10 Mbps. Note that as long as all three flows attempt to use their full fair share, the fair shares of flows 1 and 2 are less on link 2 (3.33 Mbps) than on link 1 (5 Mbps), so there will be dropping on both links. This will test the relabeling function to make sure that the incoming rates are accurately reflected on the second link. We perform two experiments (only looking at CSFQ's performance). In the first, there are three CBRs sending data at 10 Mbps each. Table 4.4 shows the average throughputs over 10 sec of the three CBR flows. As expected, these rates are closed to 3.33 Mbps. In the second experiment, we replace the three CBRs by three TCPs. Again, despite the TCP burstiness which may negatively affect the rate estimation and relabeling accuracy, each TCP gets its fair share.

4.4.7 Discussion of Simulation Results

We have tested CSFQ under a wide range of conditions; conditions purposely designed to stress its ability to achieve fair allocations. These tests, and the others we have run but cannot show here because of space limitations, suggest that CSFQ achieves a reasonable approximation of fair bandwidth allocations in most conditions. Certainly CSFQ is far superior in this regard to the status quo (FIFO or RED). Moreover, in all situations CSFQ is roughly comparable with FRED, and in some cases it achieves significantly fairer allocations. Recall that FRED requires per-packet flow classification while CSFQ does not, so we are achieving these levels of fairness in a more scalable manner. However, there is clearly room for improvement in CSFQ. We think our buffer management algorithm may not be well-tuned to the vagaries of TCP buffer usage, and so are currently looking at adopting an approach closer in spirit to RED for buffer management, while retaining the use of the labels to achieve fairness.

4.5 Related Work

An alternative to the allocation approach was recently proposed to address the problem of the ill-behaved (unfriendly) flows. This approach is called the *identification* approach and it is best exemplified by Floyd and Fall [36]. In this approach, routers use a lightweight detection algorithm to identify unfriendly flows, and then explicitly manage the bandwidth of these unfriendly flows. This bandwidth management can range from merely restricting unfriendly flows to no more than the currently highest friendly flow's share to the extreme of severely punishing unfriendly flows by dropping all of their packets.

Compared to CSFQ, the solution to implementing the identification approach described by Floyd and Fall [36] has several drawbacks. First, this solution requires routers to maintain state for each flow that has been classified as un-friendly. In contrast, CSFQ does not require core routers to maintain any per flow state. Second, designing accurate identification tests for unfriendly flows is inherently difficult. Finally, the identification approach requires that all flows to implement a similar congestion control mechanism, i.e., to be TCP friendly. We believe that this is overly restrictive as it severely limits the freedom of designing new congestion protocols that best suit application needs.

Next, we discuss in more detail the difficulty of providing accurate identification tests for unfriendly flows. One can think of the process of identifying unfriendly flows as occurring in two logically distinct stages. The first and relatively easy step is to estimate the arrival rate of a flow. The second, and harder, step is to use this arrival rate information (along with the dropping rate and other aggregate measurements) to decide if the flow is unfriendly. Assuming that friendly flows use a TCP-like adjustment method of increase-by-one and decrease-by-half, one can derive an expression (see [36] for details) for the bandwidth share S as a function of the dropping rate p, round-trip time R, and packet size B: $S \approx \frac{\gamma B}{R\sqrt{p}}$ for some constant γ. Because routers do not know the round trip time R of flows, they use the lower bound of double the propagation delay of the attached link. Unfortunately, this allows flows further away from the link to behave more aggressively without being identified as being unfriendly.

Algorithm	Simulation 1			Simulation 2	
	UDP	TCP-1	TCP-2	TCP-1	TCP-2
REDI	0.906	0.280	0.278	0.565	0.891
CSFQ	0.554	0.468	0.478	0.729	0.747

Table 4.5. Simulation 1 – The throughputs in Mbps of one UDP and two TCP flows along a 1.5 Mbps link under REDI [36] and CSFQ, respectively. Simulation 2 – The throughputs of two TCPs (where TCP-2 opens its congestion window three times faster than TCP-1), under REDI and CSFQ, respectively.

To see how this occurs in practice, consider the following two experiments using the identification algorithm described by Floyd and Fall [36], which we call RED with Identification (REDI).[5] In each case there are multiple flows traversing a 1.5 Mbps link with a latency of 3 ms; the output buffer size is 32 KB and all constants K, K_α, and K_c, respectively, are set to 400 ms. Table 4.5 shows the bandwidth allocations under REDI and CSFQ averaged over 100 sec. In the first experiment (Simulation 1), we consider a 1 Mbps UDP flow and two TCP flows; in the second experiment (Simulation 2) we have a standard TCP (TCP-1) and a modified TCP (TCP-2) that opens the congestion window three times faster. In both cases REDI fails to identify the unfriendly flow, allowing it to obtain almost two-thirds of the bandwidth. As we increase the latency of the congested link, REDI starts to identify unfriendly flows. However, for some values as high as 18 ms, it still fails to identify such flows. Thus, the identification approach still awaits a viable realization and, as of now, the allocation approach is the only demonstrated method able to deal with the problem of unfriendly flows.

---

[5] We are grateful to Sally Floyd who provided us her script implementing the REDI algorithm. We used a similar script in our simulation, with the understanding that this is a preliminary design of the identification algorithm. Our contention is that the design of such an identification algorithm is fundamentally difficult due to the uncertainty of RTT.

The problem of estimating fair-share rate has also been studied in the context of designing Available Bit Rate (ABR) algorithms for ATM networks. While the problems are similar, there are also several important differences. First, in ATM ABR, the goal is to provide explicit feedback to end systems for policing purposes so that cell loss inside the network can be prevented. In CSFQ, there is no explicit feedback and edge policing. Packets from a flow may arrive at a much higher rate than the flow's fair share rate and the goal of CSFQ is to ensure, by probabilistic dropping, that such flows do not get more service than their shares. Second, since ATM already keeps per virtual circuit (VC) state, additional per VC state is usually added to improve the accuracy and reduce the time complexity of estimating the fair share rate [20, 33, 61]. However, there are several algorithms that try to estimate the fair share without keeping per flow state [87, 95]. These algorithms rely on the flow rates communicated by the end-system. These estimates are assumed to remain accurate over multiple hops, due to the accurate explicit congestion control provided by ABR. In contrast, in CSFQ, since the number of dropped packets cannot be neglected, the flow rates are re-computed at each router, if needed (see Section 4.3.2). In addition, the estimation algorithms are themselves quite different. While the algorithms in averaging over the flow rates communicated by the end-systems, CSFQ uses linear interpolation in conjunction with exponentially averaging the traffic aggregates at the router. Our preliminary analysis and evaluation show that our estimation algorithm is more suited for our context.

4.6 Summary

In this chapter we have presented a solution that achieves fair bandwidth allocation, without requiring core routers to maintain any per-flow state. The key idea is to use the DPS technique to approximate the service provided by a reference network — in which every node implements Fair Queueing – within a SCORE network. Each node in the SCORE network implements a novel algorithm, called Core-Stateless Fair Queueing (CSFQ). With CSFQ edge routers estimate flow rates and insert them into the packet headers. Core routers simply perform probabilistic dropping on input based on these labels and an estimate of the fair share rate, the computation of which requires only aggregate measurements. Packet labels are rewritten by the core routers to reflect output rates, so this approach can handle multi-hop situations.

We have tested CSFQ and several other algorithms under a wide variety of conditions. We have found that CSFQ achieves a significant degree of fairness in all of these circumstances. While not matching the fairness benchmark of DRR, it is comparable or superior to FRED, and vastly better than the baseline cases of RED and FIFO. We know of no other approach that can achieve comparable levels of fairness without any per-flow operations in the core routers.

5 Providing Guaranteed Services in SCORE

In the previous chapter we demonstrated that using the SCORE/DPS frame-work makes it possible to implement a service with a per flow semantic (i.e., per flow isolation) in a stateless core architecture. In this chapter, we present a second example which shows that it is possible to provide per flow bandwidth and delay guarantees in a core stateless network. We achieve this goal by using the DPS technique to implement the functionalities required by the Guaranteed service [93] on both the data and control paths in a SCORE network.

The rest of this chapter is organized as follows. The next section presents the motivation behind the Guaranteed service and describes the problems with the existing solutions. Section 5.2 outlines our solution to implement the Guaranteed service in a SCORE network. Sections 5.3 and 5.4 present details of both our data and control path algorithms. Section 5.5 describes a design and a prototype implementation of the proposed algorithms in IPv4 networks. Finally, Section 5.6 describes related work, and Section 5.7 summarizes our findings.

5.1 Background

As new applications such as IP telephony, video-conferencing, audio and video streaming, and distributed games are deployed in the Internet, there is a grow-ing need to support more sophisticated services than the best-effort service. Unlike traditional applications such as file transfer, these new applications have much stricter timeliness and bandwidth requirements. For example, in order to provide a quality comparable to today's telephone service, the end-to-end delay should not exceed 100 ms [64]. Since in a global network the propagation delay alone is about 100 ms, meeting such tight delay require-ments is a challenging task [7]. Similarly, to provide high quality video and audio broadcasting, it is desirable to be able to ensure both bandwidth and delay guarantees.

To support these new applications, the IETF has proposed two service models: the Guaranteed service [93] defined in the context of Intserv [82], and the Premium service [76] defined in the context of Diffserv [32]. These

I. Stoica: Stateless Core, LNCS 2979, pp. 77-102, 2004.
© Springer-Verlag Berlin Heidelberg 2004

services have important differences in both their semantic and implementation complexity. At the service definition level, while the Guaranteed service can provide both *per flow* delay and bandwidth guarantees [93], the Premium service can provide only per flow bandwidth and *per aggregate* delay guarantees [76]. Thus, with the Premium service, if two flows have different delay requirements, say d_1 and d_2, the only way to meet both these delay requirements is to ensure a delay of $d = \min(d_1, d_2)$ to both flows. The main drawback of this approach is that it can result in very low resource utilization for the premium traffic. In particular, as we show in Appendix B.1, even if the fraction that can be allocated to the premium traffic on every link in the network is very low (e.g., 10%), the queueing delay across a large network (e.g., 15 routers) can be relative large (e.g., 240 ms). In contrast, the Guaranteed service can achieve both higher resource utilization and tighter delay bounds, by better matching flow requirements to resource usage.

At the implementation level, current solutions to provide guaranteed services require each router to process *per flow* signaling messages and maintain *per flow* state on the control path, and to perform *per flow* classification, scheduling, and buffer management on the data path. Performing per flow management inside the network affects both the network scalability and robustness. The former is because the per flow operation complexity usually increases as a function of the number of flows; the latter is because it is difficult to maintain the consistency of dynamic, and replicated per flow state in a distributed network environment. While there are several proposals that aim to reduce the number of flows inside the network, by aggregating micro-flows that follow the same path into one macro-flow [5, 49], they only alleviate this problem, but do not fundamentally solve it. This is because the number of macro flows can still be quite large in a network with many edge routers, as the number of paths is a quadratic function of the number of edge nodes.

In contrast, the Premium service is implemented in the context of the Diffserv architecture which distinguishes between edge and core routers. While edge routers process packets on a per flow or per aggregate basis, core routers do not maintain any fine grained state about the traffic; they simply forward premium packets based on a high priority bit set in the packet headers by edge routers. Pushing the complexity to the edge and maintaining a simple core makes the *data plane* highly scalable. However, the Premium service still requires admission control on the control path. One proposal is to use a centralized bandwidth broker that maintains the topology as well as the state of all nodes in the network. In this case, the admission control can be implemented by the broker, eliminating the need for maintaining distributed reservation state. However, such a centralized approach is more appropriate for an environment where most flows are long lived, and set-up and tear-down events are rare.

In summary, the Guaranteed service is more powerful but has serious limitations with respect to network scalability and robustness. On the other hand, the Premium service is more scalable, but cannot achieve the levels of flexibility and utilization of the Guaranteed service. In addition, scalable and robust admission control for the Premium service is still an open research problem.

In this chapter we show that by using the SCORE/DPS framework we can achieve the best of the two worlds: provide Guaranteed service semantic while maintaining the scalability and the robustness of the Diffserv architecture.

5.2 Solution Outline

Current solutions to implement the Guaranteed service assume a stateful network in which each router maintains per flow state. The state is used by both the admission control module in the control plane and the classifier and scheduler in the data plane.

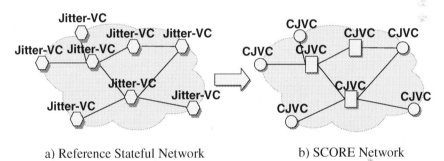

a) Reference Stateful Network   b) SCORE Network

Fig. 5.1. (a) A reference stateful network that provides the Guaranteed service [93]. Each node implements the Jitter-Virtual Clock (Jitter-VC) algorithm on the data path, and per flow admission control on the control path. (b) A SCORE network that emulates the service provided by the reference network. On the data path, each node approximates Jitter-VC with a new algorithm, called Core-Jitter Virtual Clock (CJVC). On the control path each node approximates per flow admission control.

In this chapter, we propose scheduling and admission control algorithms that provide the Guaranteed service but do not require core routers to maintain per flow state. The main idea behind our solution is to approximate a *reference* stateful network that provides the Guaranteed service in a SCORE network (see Figure 5.1). The key technique used to implement these algorithms is Dynamic Packet State (DPS). On the control path, our solution aims to emulate per flow admission control, while on the data path, our algorithm aims to emulate a reference network in which every node implements the Delay-Jitter-Controlled Virtual Clock (Jitter-VC) algorithm.

Among many scheduling disciplines that can implement the Guaranteed service we chose Jitter-VC for several reasons. First, unlike various Fair

Queueing algorithms [31, 79], in which a packet's deadline can depend on state variables of *all* active flows, in Jitter-VC a packet's deadline depends only on the state variables of the flow it belongs to. This property of Jitter-VC makes the algorithm easier to approximate in a SCORE network. In particular, the fact that packet's deadline can be computed exclusively based on the state variables of the flow it belongs to, makes it possible to eliminate the need to replicate and maintain per flow state at all nodes along the path. Instead, per flow state can be stored only at the ingress node, inserted into the packet header by the ingress node, and retrieved later by core nodes, which then use it to determine the packet's deadline. Second, by regulating traffic inside the network using delay-jitter-controllers (discussed below), it can be shown that with very high probability, the number of packets in the server at any given time is significantly smaller than the number of flows (see Section 5.3.3). This helps to simplify the scheduler.

In the next section, we present techniques that eliminate the need for data plane algorithms to use per flow state at core nodes. In particular, at core nodes, packet classification is no longer needed and packet scheduling is based on the state carried in packet headers, rather than per flow state stored locally at each node. In Section 5.4, we will show that fully distributed admission control can also be achieved without the need for maintaining per flow state at core nodes.

5.3 Data Plane: Scheduling without Per Flow State

In this section, we first describe Jitter-VC, which is used to achieve guaranteed services in the reference network, and then present our algorithm, called Core-Jitter-VC (CJVC). CJVC uses the Dynamic Packet State (DPS) technique to emulate Jitter-VC in a SCORE network. In Appendix B.3 we present an analysis to show that a network of routers implementing CJVC provides the same delay bound as a network of routers implementing the Jitter-VC algorithm.

5.3.1 Jitter Virtual Clock (Jitter-VC)

Jitter-VC is a non-work-conserving version of the Virtual Clock algorithm [127]. It uses a combination of a delay-jitter rate-controller [112, 126] and a Virtual Clock scheduler. The algorithm works as follows: each packet is assigned an eligible time and a deadline upon its arrival. The packet is held in the rate-controller until it becomes eligible, i.e., the system time exceeds the packet's eligible time (see Figure 5.2(a)). The scheduler then orders the transmission of eligible packets according to their deadlines.

For the kth packet of flow i, its eligible time, $e_{i,j}^k$, and deadline, $d_{i,j}^k$, at the jth node on its path are computed as follows:

Notation	Comments
p_i^k	the k-th packet of flow i
l_i^k	length of p_i^k
$a_{i,j}^k$	arrival time of p_i^k at node j
$s_{i,j}^k$	time when p_i^k was sent by node j
$e_{i,j}^k$	eligible time of p_i^k at node j
$d_{i,j}^k$	deadline of p_i^k at node j
$g_{i,j}^k$	time ahead of schedule: $g_{i,j}^k = d_{i,j}^k + \tau_j - s_{i,j}^k$
δ_i^k	slack delay of p_i^k
π_j	propagation delay between nodes j and $j+1$
τ_j	transmission time of a maximum size packet at node j

Table 5.1. Notations used in Section 5.3.

$$e_{i,j}^1 = a_{i,j}^1$$
$$e_{i,j}^k = \max(a_{i,j}^k + g_{i,j-1}^k, d_{i,j}^{k-1}), \quad i,j \geq 1, k > 1 \tag{5.1}$$
$$d_{i,j}^k = e_{i,j}^k + \frac{l_i^k}{r_i}, \quad i,j,k \geq 1 \tag{5.2}$$

where l_i^k is the length of the packet, r_i is the reserved rate for the flow, $a_{i,j}^k$ is the packet's arrival time at the jth node traversed by the packet, and $g_{i,j}^k$, stamped into the packet header by the previous node, is the amount of time the packet was transmitted before its schedule, i.e., the difference between the packet's deadline and its actual departure time at node $j-1$. To account for the fact that the packet transmission is not preemptive, and as a result a packet can miss its deadline by the time it takes to transmit a packet of maximum size, τ_j [127], we inflate the packet delay by τ_j when computing $g_{i,j}^k$ (see Table 5.1).

Intuitively, the algorithm eliminates the delay variation of different packets by forcing all packets to incur the maximum allowable delay. The purpose of having $g_{i,j-1}^k$ is to compensate at node j the variation of delay due to load fluctuation at the previous node, $j-1$. Such regulations limit the traffic burstiness caused by network load fluctuations, and as a consequence, reduce both buffer space requirements and the scheduler complexity.

It has been shown that if a flow's long term arrival rate is no greater than its reserved rate, a network of Virtual Clock servers can provide the same delay guarantee to the flow as a network of Weighted Fair Queueing (WFQ) servers [35, 47, 100]. In addition, it has been shown that a network of Jitter-VC servers can provide the same delay guarantees as a network of Virtual Clock servers [28, 42]. Therefore, a network of Jitter-VC servers can provide the same guaranteed service as a network of WFQ servers.

5.3.2 Core-Jitter-VC (CJVC)

In this section we propose a variant of Jitter-VC, called Core-Jitter-VC (CJVC), which does not require per flow state at core nodes. In addition, we show that a network of CJVC servers can provide the same guaranteed service as a network of Jitter-VC servers.

CJVC uses the DPS technique. The key idea is to have the ingress node to encode scheduling parameters in each packet's header. The core routers can then make scheduling decisions based on the parameters encoded in packet headers, thus eliminating the need for maintaining per flow state at core nodes. As suggested by Eqs. (5.1) and (5.2), the Jitter-VC algorithm needs two state variables for each flow i: r_i, which is the reserved rate for flow i and $d_{i,j}^k$, which is the deadline of the last packet from flow i that was served by node j. While it is straightforward to eliminate r_i by putting it in the packet header, it is not trivial to eliminate $d_{i,j}^k$. The difference between r_i and $d_{i,j}^k$ is that while all nodes along the path keep the same r_i value for flow i, $d_{i,j}^k$ is a dynamic value that is computed iteratively at each node. In fact, the eligible time and the deadline of p_i^k depend on the deadline of the previous packet of the same flow, i.e., $d_{i,j}^{k-1}$.

A naive implementation using the DPS technique would be to precompute the eligible times and the deadlines of the packet at all nodes along its path and insert all of them in the header. This would eliminate the need for core nodes to maintain $d_{i,j}^k$. The main disadvantage of this approach is that the amount of information carried by the packet increases with the number of hops along the path. The challenge then is to design algorithms that compute $d_{i,j}^k$ for all nodes while requiring a minimum amount of state in the packet header.

Notice that in Eq. (5.1), the reason for node j to maintain $d_{i,j}^k$ is that it will be used to compute the deadline and the eligible time of the next packet. Since it is only used in a `max` operation, we can eliminate the need for $d_{i,j}^k$ if we can ensure that the other term in `max` is never less than $d_{i,j}^k$. The key idea is then to use a *slack* variable associated with each packet, denoted δ_i^k, such that for every core node j along the path, the following holds

$$a_{i,j}^k + g_{i,j-1}^k + \delta_i^k \geq d_{i,j}^{k-1}, \quad j > 1 \tag{5.3}$$

By replacing the first term of `max` in Eq. (5.1) with $a_{i,j}^k + g_{i,j-1}^k + \delta_i^k$, the computation of the eligible time reduces to

$$e_{i,j}^k = a_{i,j}^k + g_{i,j-1}^k + \delta_i^k, \quad j > 1 \tag{5.4}$$

Therefore, by using one additional DPS variable, δ_i^k, we eliminate the need for maintaining $d_{i,j}^k$ at the core nodes.

The derivation of δ_i^k proceeds in two steps. First, we express the eligible time of packet p_i^k at an arbitrary core node j, $e_{i,j}^k$, as a function of the eligible

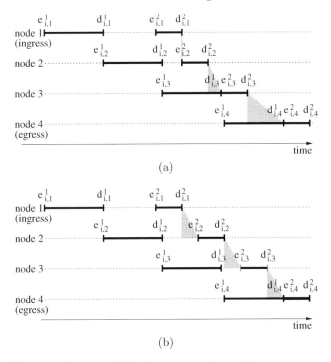

Fig. 5.2. The time diagram of the first two packets of flow i along a four node path under (a) Jitter-VC, and (b) CJVC, respectively. Propagation times, π_j, and transmission times of maximum size packets, τ_j, are ignored.

time of p_i^k at the ingress node, $e_{i,1}^k$, (see Eq. (5.7)). Second, we use this result and Eq. (5.4) to derive a lower bound for δ_i^k.

We now proceed with the first step. Recall that $g_{i,j-1}^k$ represents the time by which p_i^k is transmitted before its schedule at node $j-1$, i.e., $d_{i,j-1}^k + \tau_{j-1} - s_{i,j-1}^k$, where τ_{j-1} is the maximum time by which a packet can miss its deadline at node $j-1$. Let π_{j-1} denote the propagation delay between nodes $j-1$ and j. Then the arrival time of p_i^k at node j, $a_{i,j}^k$, is given by

$$a_{i,j}^k = s_{i,j-1}^k + \pi_{j-1} + \tau_{j-1} \qquad (5.5)$$
$$= d_{i,j-1}^k - g_{i,j-1}^k + \pi_{j-1} + \tau_{j-1}.$$

By replacing $a_{i,j}^k$, given by the above expression, in Eq. (5.4), and then using Eq. (5.2), we obtain

$$e_{i,j}^k = d_{i,j-1}^k + \delta_i^k + \pi_{j-1} + \tau_{j-1} \qquad (5.6)$$
$$= e_{i,j-1}^k + \frac{l_i^k}{r_i} + \delta_i^k + \pi_{j-1} + \tau_{j-1}.$$

By iterating over the above equation we express $e_{i,j}^k$ as a function of $e_{i,1}^k$:

$$e_{i,j}^k = e_{i,1}^k + (j-1)\left(\frac{l_i^k}{r_i} + \delta_i^k\right) + \sum_{m=1}^{j-1}(\pi_m + \tau_m), \quad j > 1 \qquad (5.7)$$

We are now ready to compute δ_i^k. Recall that the goal is to compute the minimum δ_i^k which ensures that Eq. (5.3) holds for every node along the path. After combining Eq. (5.3), Eq. (5.4) and Eq. (5.2) this reduces to ensure that

$$e_{i,j}^k \geq d_{i,j}^{k-1} \Rightarrow e_{i,j}^k \geq e_{i,j}^{k-1} + \frac{l_i^{k-1}}{r_i}, \quad j > 1 \qquad (5.8)$$

By plugging $e_{i,j}^k$ and $e_{i,j}^{k-1}$ as expressed by Eq. (5.7) into Eq. (5.8), we get

$$\delta_i^k \geq \delta_i^{k-1} + \frac{l_i^{k-1} - l_i^k}{r_i} + \frac{e_{i,1}^{k-1} + l_i^{k-1}/r_i - e_{i,1}^k}{(j-1)}, \quad j > 1 \qquad (5.9)$$

From Eqs. (5.1) and (5.2) we have $e_{i,1}^k \geq d_{i,1}^{k-1} = e_{i,1}^{k-1} + l_i^{k-1}/r_i$. Thus, the right-hand side term in Eq. (5.9) is maximized when $j = h$. As a result we compute δ_i^k as

$$\delta_i^1 = 0, \qquad (5.10)$$

$$\delta_i^k = \max\left(0, \delta_i^{k-1} + \frac{l_i^{k-1} - l_i^k}{r_i} - \frac{e_{i,1}^k - e_{i,1}^{k-1} - l_i^{k-1}/r_i}{h-1}\right), \quad k > 1, h > 1.$$

In this way, CJVC ensures that the eligible time of every packet, p_i^k, at node j is no smaller than the deadline of the previous packet of the same flow at node j, i.e., $e_{i,j}^k \geq d_{i,j}^{k-1}$. In addition, the Virtual Clock scheduler ensures that the deadline of every packet is not missed by more than τ_j [127].

In Appendix B.2, we have shown that a network of CJVC servers provides the same worst case delay bounds as a network of Jitter-VC servers. More precisely, we have proven the following property.

Theorem 2 *The deadline of a packet at the last hop in a network of CJVC servers is equal to the deadline of the same packet in a corresponding network of Jitter-VC servers.*

The example in Figure 5.2 provides some intuition behind the above property. The basic observation is that, with Jitter-VC, not counting the propagation delay, the difference between the eligible time of packet p_i^k at node j and its deadline at the previous node, $j - 1$, i.e., $e_{i,j}^k - d_{i,j-1}^k$, never *decreases* as the packet propagates along the path. Consider the second packet in Figure 5.2. With Jitter-VC, the differences $e_{i,j}^2 - d_{i,j-1}^2$ (represented by the bases of the gray triangles) increase in j. By introducing the slack variable δ_i^k, CJVC *equalizes* these delays. While this change may increase the delay

---

ingress node

---

on packet p arrival
   $i = get\_flow(p)$;
   if (first_packet_of_flow(p, i))
      $e_i = current\_time$;
      $\delta_i = 0$;
   else
      $\delta_i = \max(0, \delta_i + (l_i - length(p))/r_i -$
          $\max(current\_time - d_i, 0)/(h-1))$; /* Eq. (5.10) */
      $e_i = \max(current\_time, d_i)$;
   $l_i = length(p)$;
   $d_i = e_i + l_i/r_i$;
on packet p transmission
   $label(p) \leftarrow (r_i, d_i - current\_time, \delta_i)$;

---

core/egress node

---

  on packet p arrival
    $(r, g, \delta) \leftarrow label(p)$;
    $e = current\_time + g + \delta$; /* Eq. (5.4) */
    $d = e + length(p)/r$
  on packet p transmission
    if (core node)
       $label(p) \leftarrow (r, d - current\_time, \delta)$;
    else /* this is an egress node */
       $clear\_label(p)$;

Fig. 5.3. Algorithms performed by ingress, core, and egress nodes at the packet arrival and departure. Note that core and egress nodes do not maintain per flow state.

of the packet at intermediate hops, it does not affect the end-to-end delay bound.

Figure 5.3 shows the computation of the scheduling parameters $e_{i,j}^k$ and $d_{i,j}^k$ by a CJVC server. The number of hops h is computed at the admission time as discussed in Section 5.4.1.

5.3.3 Data Path Complexity

While our algorithms do not maintain per flow state at core nodes, there is still the need for core nodes to perform regulation and packet scheduling based on eligible times and deadlines. The natural question to ask is: why is this a more scalable scheme than previous solutions requiring per flow management?

There are several scalability bottlenecks for solutions requiring per flow management. On the data path, the expensive operations are per flow classification and scheduling. On the control path, complexity results from the maintenance of consistent and dynamic state in a distributed environment.

Among the three, it is easiest to reduce the complexity of the scheduling algorithm as there is a natural tradeoff between the complexity and the flexibility of the scheduler [119]. In fact, a number of techniques have already been proposed to reduce scheduling complexity, including those requiring constant time complexity [98, 120, 125].

We also note that due to the way we regulate traffic, it can be shown that with very high probability, the number of packets in the server at any given time is significantly smaller than the number of flows. This will further reduce the scheduling complexity and in addition reduce the buffer space requirement. More precisely, in Appendix C we prove the following result.

Theorem 3 *Consider a server traversed by n flows. Assume that the arrival times of the packets from different flows are independent, and that all packets have the same size. Then, for any given probability ε, the queue size at any instant during a server busy period is asymptotically bounded above by s, where*

$$s = \sqrt{\beta n \left(\frac{\ln n}{2} - \frac{\ln \varepsilon}{2} - 1 \right)}, \tag{5.11}$$

with a probability larger than $1 - \varepsilon$. For identical reservations $\beta = 1$; for heterogeneous reservations $\beta = 3$.

As an example, let $n = 10^6$, and $\varepsilon = 10^{-10}$, which is the same order of magnitude as the probability of a packet being corrupted at the physical layer. Then, by Eq. (5.11) we obtain $s = 4174$ if all flows have identical reservations, and $s = 7230$ if flows have heterogeneous reservations. Thus the probability of having more packets in the queue than specified by Eq. (5.11) can be neglected at the level of the entire system even in the context of guaranteed services.

# flows (n)	bound (s)	max. queue size	# flows (n)	bound (s)	max. queue size
100	31	28	100	53	30
1,000	109	100	1,000	188	95
10,000	374	284	10,000	648	309
100,000	1276	880	100,000	2210	904
1,000,000	4310	2900	1,000,000	7465	2944
(a)			(b)		

Table 5.2. The upper bound of the queue size, s, computed by Eq. (5.11) for $\varepsilon = \frac{10^{-5}}{n}$ (where n is the number of flows) versus the maximum queue size achieved during the first n time slots of a busy period over 10^5 independent trials, during the first n time slots of a busy period: (a) when all flows have identical reservations; (b) when the flows' reservations differ by a factor of 20.

In Table 5.2 we compare the bounds given by Eq. (5.11) to simulation results. In each case we report the maximum queue size achieved during the first n time slots of a busy period over 10^5 independent trials. We note that in the case of all flows having identical reservations we are guaranteed that if the queue does not overflow during the first n time slots of a busy period, it will not overflow during the rest of the busy period (see Corollary 1). Since the probability that a buffer will overflow during the first n time slots is no larger than n times the probability of the buffer overflowing during an arbitrary time slot, we use $\varepsilon = \frac{10^{-5}}{n}$ to compute the corresponding bounds.[1]

The results show that our bounds are reasonably close (within a factor of two) when all reservations are identical, but are more conservative when the reservations are different. Finally, we make three comments. First, by performing per packet regulation at every core node, the bounds given by Eq. (5.11) hold for any core node and are *independent* of the path length. Second, if the flows' arrival patterns are not independent, we can easily enforce this by randomly delaying the first packet from each backlogged period of the flow at ingress nodes. This will increase the end-to-end packet delay by at most the queueing delay of one extra hop. Third, the bounds given by Eq. (5.11) are asymptotic. In particular, in proving the results in Appendix C we make the assumption that $n \gg s$. However, this a reasonable assumption in practice, as the most interesting cases involve high values for n, and, as suggested by Eq. (5.11) and the results in Table 5.2, even for small values of ε (e.g., 10^{-10}), n is much larger than s.

5.4 Control Plane: Admission Control with no Per Flow State

A key component of any architecture that provides guaranteed services is the admission control. The main job of the admission control is to ensure that the network resources are not over-committed. In particular it has to ensure that the sum of the reservation rates of all flows that traverse any link in the network is no larger than the link capacity, i.e., $\sum_i r_i < C$. A new reservation request is granted if it passes the admission test at each hop along its path. As discussed in this chapter's introduction, implementing such a functionality is not trivial: traditional distributed architectures based on signaling protocols are not scalable and are less robust due to the requirement of maintaining dynamic and replicated state; centralized architectures have scalability and availability concerns.

In this section, we propose a fully distributed architecture for implementing admission control. Like most distributed admission control architectures,

---

[1] More formally, let ε' be the probability that the buffer does not overflow during the first n time slots of the busy period. Then by taking $\varepsilon' = n \cdot \varepsilon$, Eq. (5.11) becomes $s = \sqrt{\beta n (\ln n - (\ln \varepsilon')/2 - 1)}$.

in our solution, each node keeps track of the aggregate reservation rate for each of its out-going links and makes local admission control decisions. However, unlike existing reservation protocols, this distributed admission control process is achieved without core nodes maintaining per flow state.

5.4.1 Ingress-to-Egress Admission Control

We consider an architecture in which a lightweight signaling protocol is used within the SCORE domain. Edge routers are the interface between this signaling protocol and an inter-domain signaling protocol such as RSVP. For the purpose of this discussion, we consider only unicast reservations. In addition, we assume a mechanism like the one proposed by Stoica and Zhang [103] or Multi-Protocol Label Switching (MPLS) [17] that can be used to pin a flow to a route.

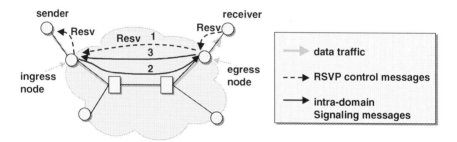

Fig. 5.4. Ingress-egress admission control when RSVP is used outside the SCORE domain.

From the point of view of RSVP, a path through the SCORE domain is just a virtual link. There are two basic control messages in RSVP: *Path* and *Resv*. These messages are processed only by edge nodes; no operations are performed inside the domain. For the ingress node, upon receiving a *Path* message, it simply forwards it through the domain. For the egress node, upon receiving the first *Resv* message for a flow (i.e., there was no RSVP state for the flow at the egress node before receiving the message), it will forward the message (message "1" in Figure 5.4) to the corresponding ingress node, which in turn will send a special signaling message (message "2" in Figure 5.4) along the path toward the egress node. Upon receiving the signaling message, each node along the path performs a local admission control test as described in Section 5.4.2. In addition, the message carries a counter, h, that is incremented at each hop. The final value h is used for computing the slack delay, δ, (see Eq. (5.10)). If we use the route pinning mechanism described in Chapter 6, message "2" is also used to compute the label of the path between the ingress and egress. This label is used then by the ingress node to make sure that all data packets of the flow are forwarded along the same path. When

Notation	Comments
r_i	flow i's reserved rate
b_i^k	total number of bits flow i is entitled to transmit during $[s_{i,1}^{k-1}, s_{i,1}^k]$, i.e., $b_i^k = r_i(s_{i,1}^k - s_{i,1}^{k-1})$
$R(t)$	aggregate reservation at time t
$R_{bound}(t)$	upper bound of $R(t)$, used by admission test
$R_{DPS}(t)$	estimate of $R(t)$, computed by using DPS
$R_{new}(t)$	sum of all new reservations accepted from the beginning of current estimation interval until t
$R_{cal}(t)$	upper bound of $R(t)$, used to calibrate R_{bound}, computed based on R_{DPS} and R_{new}

Table 5.3. Notations used in Section 5.4.3.

the signaling message "2" reaches the egress node, it is reflected back to the sender, which makes the final decision (message "3" in Figure 5.4). RSVP refresh messages for a flow that already has per flow RSVP state installed at edge routers will not trigger additional signaling messages inside the domain.

Since RSVP uses raw IP or UDP to send control messages, there is no need for retransmission for our signaling messages, as message loss will not break the RSVP semantics. If the sender does not receive a reply after a certain timeout, it simply drops the *Resv* message. In addition, as we will show in Section 5.4.3, there is no need for a special termination message inside the domain when a flow is torn down.

5.4.2 Per-Hop Admission Control

Each node needs to ensure that $\sum_i r_i < C$ holds at all times. At first sight, one simple solution that implements this test and also avoids per flow state is for each node to maintain the aggregate reserved rate, R, where R is updated to $R = R + r$ when a new flow with the reservation rate r is admitted, and to $R = R - r'$ when a flow with the reservation rate r' terminates. The admission control then reduces to checking whether $R + r \le C$ holds. However, it can be easily shown that such a simple solution is not robust with respect to various failure conditions such as packet loss, partial reservation failures, and network node crashes. To handle packet loss, when a node receives a set-up or tear-down message, the node has to be able to tell whether it is a duplicate of a message already processed. To handle partial reservation failures, a node needs to "remember" what decision it made for the flow in a previous pass. That is why all existing solutions maintain per flow reservation state, be it hard state as in ATM UNI or soft state as in RSVP. However, maintaining *consistent* and *dynamic* state in a *distributed* environment is itself challenging. Fundamentally, this is because the update operations assume a *transaction* semantic, which is difficult to implement in a distributed environment [4, 117].

In the remaining of the section, we show that by using DPS, it is possible to significantly reduce the complexity of admission control in a distributed environment. Before we present the details of the algorithm, we point out that our goal is to estimate a close *upper bound* on the aggregate *reserved* rate. By using this bound in the admission test we avoid over-provisioning, which is a necessary condition to provide deterministic service guarantees. This is in contrast to many measurement-based admission control algorithms [62, 110], which, in the context of supporting controlled load or statistical services, base their admission test on the measurement of the *actual* amount of traffic transmitted. To achieve this goal, our algorithm uses two techniques. First, a conservative upper bound of R, denoted R_{bound}, is maintained at each core node and is used for making admission control decisions. R_{bound} is updated with a simple rule: $R_{bound} = R_{bound} + r$ whenever a new request of a rate r is accepted. It should be noted that in order to maintain the invariant that R_{bound} is an upper bound of R, this algorithm does not need to detect duplicate request messages, generated either due to retransmission in case of packet loss or retry in case of partial reservation failures. Of course, the obvious problem with this algorithm is that R_{bound} will diverge from R. In the limit, when R_{bound} reaches the link capacity C, no new requests can be accepted even though there might be available capacity.

To address this problem, a separate algorithm is introduced to periodically estimate the aggregate reserved rate. Based on this estimate, a second upper bound for R, denoted R_{cal}, is computed and used to recalibrate R_{bound}. An important aspect of the estimation algorithm is that the discrepancy between the upper bound R_{cal} and the actual reserved rate, R, can be bounded. The recalibration then becomes the choice between the minimum of the two upper bounds R_{bound} and R_{cal}. The estimation algorithm is based on DPS and does not require core routers to maintain per flow state.

Our algorithms have several important properties. First, they are robust in the presence of network losses and partial reservation failures. Second, while they can over-estimate R, they will never underestimate R. This ensures the semantics of the guaranteed service – while over-estimation can lead to under-utilization of network resources, under-estimation can result in over-provisioning and violation of performance guarantees. Finally, the proposed estimation algorithms are self-correcting in the sense that over-estimation in a previous period will be corrected in the next period. This greatly reduces the possibility of serious resource under-utilization.

5.4.3 Aggregate Reservation Estimation Algorithm

In this section, we present the estimation algorithm of the aggregate reserved rate which is performed at each core node. In particular, we will describe how R_{cal} is computed and how it is used to recalibrate R_{bound}. In designing the algorithm for computing R_{cal}, we want to balance between two goals: (a)

R_{cal} should be an upper bound on R; (b) over-estimation errors should be corrected and kept to the minimum.

To compute R_{cal}, we start with an inaccurate estimate of R, denoted R_{DPS}, and then make adjustments to account for estimation inaccuracies. In the following, we first present the algorithm that computes R_{DPS}, then describe the possible inaccuracies and the corresponding adjustment algorithms.

The estimate R_{DPS} is calculated using the DPS technique: ingress nodes insert additional state in packet headers, state which is in turn used by core nodes to estimate the aggregate reservation R. In particular, a new state, b_i^k, is inserted in the header of packet p_i^k:

$$b_i^k = r_i(s_{i,1}^k - s_{i,1}^{k-1}), \qquad (5.12)$$

where $s_{i,1}^{k-1}$ and $s_{i,1}^k$ are the times the packets p_i^{k-1} and p_i^k are transmitted by the ingress node. Therefore, b_i^k represents the total amount of bits that flow i is entitled to send during the interval $[s_{i,1}^{k-1}, s_{i,1}^k]$. The computation of R_{DPS} is based on the following simple observation: the sum of b values of all packets of flow i during an interval is a good approximation for the total number of bits that flow i is *entitled* to send during that interval according to its reserved rate. Similarly, the sum of b values of *all* packets is a good approximation for the total number of bits that all flows are entitled to send during the corresponding interval. Dividing this sum by the length of the interval gives the aggregate reservation rate. More precisely, let us divide time into intervals of length T_W: $(u_k, u_{k+1}]$, $k > 0$. Let $b_i(u_k, u_{k+1})$ be the sum of b values of packets in flow i received during $(u_k, u_{k+1}]$, and let $B(u_k, u_{k+1})$ be the sum of b values of *all* packets during $(u_k, u_{k+1}]$. The estimate is then computed at the end of each interval $(u_k, u_{k+1}]$ as follows

$$R_{DPS}(u_{k+1}) = \frac{B(u_k, u_{k+1})}{u_{k+1} - u_k} = \frac{B(u_k, u_{k+1})}{T_W}. \qquad (5.13)$$

While simple, the above algorithm may introduce two types of inaccuracies. First, it ignores the effects of the delay jitter and the packet inter-departure times. Second, it does not consider the effects of accepting or terminating a reservation in the middle of an estimation interval. In particular, having newly accepted flows in the interval may result in the under-estimation of $R(t)$ by $R_{DPS}(t)$. To illustrate this, consider the following simple example: there are no guaranteed flows on a link until a new request with rate r is accepted at the end of an estimation interval $(u_k, u_{k+1}]$. If no data packet from the new flow reaches the node before u_{k+1}, $B(u_k, u_{k+1})$ would be 0, and so would be $R_{DPS}(u_{k+1})$. However, the correct value should be r.

In the following, we present the algorithm to compute an upper bound of $R(u_{k+1})$, denoted $R_{cal}(u_{k+1})$. In doing this we account for both types of inaccuracies. Let $\mathcal{L}(t)$ denote the set of reservations at time t. Our goal is then to

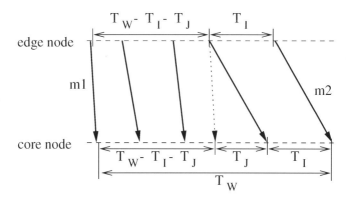

Fig. 5.5. The scenario in which the lower bound of b_i, i.e., $r_i(T_W - T_I - T_J)$, is achieved. The arrows represent packet transmissions. T_W is the averaging window size; T_I is an upper bound on the packet inter-departure time; T_J is an upper bound on the delay jitter. Both $m1$ and $m2$ miss the estimation interval T_W.

bound the aggregate reservation at time u_{k+1}, i.e., $R(u_{k+1}) = \sum_{i \in \mathcal{L}(u_{k+1})} r_i$. Consider the division of $\mathcal{L}(u_{k+1})$ into two subsets: the subset of new reservations that were accepted during the interval $(u_k, u_{k+1}]$, denoted $\mathcal{N}(u_{k+1})$, and the subset containing the rest of reservations which were accepted no later than u_{k+1}. Next, we express $R(u_{k+1})$ as

$$R(u_{k+1}) = \sum_{i \in \mathcal{L}(u_{k+1}) \backslash \mathcal{N}(u_{k+1})} r_i + \sum_{i \in \mathcal{N}(u_{k+1})} r_i. \qquad (5.14)$$

The idea is then to derive an upper bound for each of the two right-hand side terms, and compute R_{cal} as the sum of these two bounds. To bound $\sum_{i \in \mathcal{L}(u_{k+1}) \backslash \mathcal{N}(u_{k+1})} r_i$, we note that

$$B(u_k, u_{k+1}) \geq \sum_{i \in \mathcal{L}(u_{k+1}) \backslash \mathcal{N}(u_{k+1})} b_i(u_k, u_{k+1}). \qquad (5.15)$$

The reason that (5.15) is an inequality instead of an equality is that when there are flows terminating during the interval $(u_k, u_{k+1}]$, their packets may still have contributed to $B(u_k, u_{k+1})$ even though they do not belong to $\mathcal{L}(u_{k+1}) \backslash \mathcal{N}(u_{k+1})$. Next, we compute a lower bound for $b_i(u_k, u_{k+1})$. By definition, since $i \in \mathcal{L}(u_{k+1}) \backslash \mathcal{N}(u_{k+1})$, it follows that flow i holds a reservation during the entire interval $(u_k, u_{k+1}]$. Let T_I be the maximum inter-departure time between two consecutive packets of a flow at the edge node, and let T_J be the maximum delay jitter of a flow. T_J represents the maximum difference between the delays experienced by two packets between ingress and egress nodes. The ingress-egress delay of a packet represents the difference between the arrival time of the packet at the egress node and the departure time of the packet at the ingress node. In the remainder of this section, we assume that T_W is chosen such that both T_I and T_J are much smaller than T_W. Now, consider the scenario shown in Figure 5.5 in which a core node receives the packets $m1$ and $m2$ just outside the estimation window. Assuming the worst case in which $m1$ incurs the lowest possible delay, $m2$ incurs the maximum possible delay, and that the last packet before $m2$ departs T_I seconds earlier, it is easy to see that that the sum of the b values carried by the packets received during the estimation interval by the core node cannot be smaller than $r_i(T_W - T_I - T_J)$. Thus, we have

$$b_i(u_k, u_{k+1}) > r_i(T_W - T_I - T_J), \tag{5.16}$$
$$\forall i \in \mathcal{L}(u_{k+1}) \setminus \mathcal{N}(u_{k+1}). \tag{5.17}$$

By combining Eqs. (5.15) and (5.16), and Eq. (5.13) we obtain

$$\sum_{i \in \mathcal{L}(u_{k+1}) \setminus \mathcal{N}(u_{k+1})} r_i < \sum_{i \in \mathcal{L}(u_{k+1}) \setminus \mathcal{N}(u_{k+1})} \frac{b_i(u_k, u_{k+1})}{T_W(1 - f)}$$
$$\leq \frac{R_{DPS}(u_{k+1})}{1 - f}, \tag{5.18}$$

where $f = (T_I + T_J)/T_W$.

Next, we bound the second right-hand side term in Eq. (5.14): $\sum_{i \in \mathcal{N}(u_{k+1})} r_i$ For this, we introduce a new global variable R_{new}. R_{new} is initialized at the beginning of each interval $(u_k, u_{k+1}]$ to zero, and is updated to $R_{new} + r$ every time a new reservation, r, is accepted. Let $R_{new}(t)$ denote the value of this variable at time t. For simplicity, here we assume that a flow which is granted a reservation during the interval $(u_k, u_{k+1}]$ becomes active no later than u_{k+1}.[2] Then it is easy to see that

$$\sum_{i \in \mathcal{N}(u_{k+1})} r_i \leq R_{new}(u_{k+1}). \tag{5.19}$$

Eq. (5.19) holds when no duplicate reservation requests are processed, and none of the new accepted reservations terminate during the interval. Then we define $R_{cal}(u_{k+1})$ as

$$R_{cal}(u_{k+1}) = \frac{R_{DPS}(u_{k+1})}{1 - f} + R_{new}(u_{k+1}). \tag{5.20}$$

From Eq. (5.14), and Eqs. (5.18) and (5.19), it follows easily that $R_{cal}(u_{k+1})$ is an upper bound for $R(u_{k+1})$, i.e., $R_{cal}(u_{k+1}) > R(u_{k+1})$. Finally, we use $R_{cal}(u_{k+1})$ to recalibrate the upper bound of the aggregate reservation, R_{bound}, at u_{k+1} as

$$R_{bound}(u_{k+1}) = \min(R_{bound}(u_k), R_{cal}(u_{k+1})). \tag{5.21}$$

Per-hop Admission Control

---

on reservation request r
if $(R_{bound} + r \leq C)$ /* *perform admission test* */
    $R_{new} = R_{new} + r;$
    $R_{bound} = R_{bound} + r;$
    accept request;
else
    deny request;
on reservation termination r /* *optional* */
    $R_{bound} = R_{bound} - r;$

Aggregate Reservation Bound Comp.

---

on packet arrival p
    $b \leftarrow get\_b(p);$ /* *get b value inserted by ingress (Eq. (5.12))* */
    $L = L + b;$
on time-out T_W
    $R_{DPS} = L/T_W;$ /* *estimate aggregate reservation* */
    $R_{bound} = \min(R_{bound}, R_{DPS}/(1-f) + R_{new});$
    $R_{new} = 0;$

Fig. 5.6. The control path algorithms executed by core nodes; R_{new} is initialized to 0.

Figure 5.6 shows the pseudocode of the control algorithms at core nodes. Next we make several observations.

First, the estimation algorithm uses only the information in the current interval. This makes the algorithm robust with respect to loss and duplication of signaling packets since their effects are "forgotten" after one time interval. As an example, if a node processes both the original and a duplicate of the same reservation request during the interval $(u_k, u_{k+1}]$, R_{bound} will be updated twice for the same flow. However, this erroneous update will not be reflected in the computation of $R_{DPS}(u_{k+2})$, since its computation is based only on the b values received during $(u_{k+1}, u_{k+2}]$.

As a consequence, an important property of our admission control algorithm is that it can asymptotically reach a link utilization of $C(1-f)/(1+f)$. In particular, the following property is proven in Appendix B.4:

Theorem 4 *Consider a link of capacity C at time t. Assume that no reservation terminates and there are no reservation failures or request losses after time t. Then if there is sufficient demand after t the link utilization approaches asymptotically $C(1-f)/(1+f)$.*

---

[2]Otherwise, to account for the case in which a reservation accepted during the interval $(u_{k-1}, u_k]$ becomes active after $u_k + RTT$, we need to subtract $RTT \times R_{new}(u_k)$ from $B(u_k, u_{k+1})$.

Second, note that since $R_{cal}(u_k)$ is an upper bound of $R(u_k)$, a simple solution would be to use $R_{cal}(u_k) + R_{new}$, instead of R_{bound}, to perform the admission test during $(u_k, u_{k+1}]$. The problem with this approach is that R_{cal} can overestimate the aggregate reservation R. An example is given in Section 5.5 to illustrate this issue (Figure 5.10(b)).

Third, we note that a possible optimization of the admission control algorithm is to add reservation termination messages (see Figure 5.6). This will reduce the discrepancy between the upper bound, R_{bound}, and the aggregate reservation R. However, in order to guarantee that R_{bound} remains an upper bound for R, we need to ensure that a termination message is sent at most *once*, i.e., there are no retransmissions if the message is lost. In practice, this property can be enforced by edge nodes, which maintain per flow state.

Finally, to ensure that the maximum inter-departure time is no larger than T_I, the ingress node may need to send a dummy packet in the case when no data packet arrives for a flow during an interval T_I. This can be achieved by having the ingress node maintain a timer with each flow. An optimization would be to aggregate all "micro-flows" between each pair of ingress and egress nodes into one flow, compute b values based on the aggregated reservation rate, and insert a dummy packet only if there is no data packet of the aggregate flow during an interval.

5.5 Experimental Results

We have fully implemented the algorithms described in this chapter in FreeBSD v2.2.6 and deployed them in a testbed consisting of 266 MHz and 300 MHz Pentium II PCs connected by point-to-point 100 Mbps Ethernets. The testbed allows the configuration of a path with up to two core routers. The details of the implementations and of the state encoding are presented in Chapter 8.

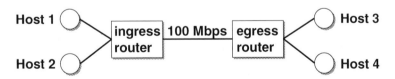

Fig. 5.7. The test configuration used in experiments.

In the remainder of this section, we present results from four simple experiments. The experiments are designed to illustrate the microscopic behaviors of the algorithms, rather than their scalability. All experiments were run on the topology shown in Figure 5.7. The first router is configured as an ingress node, while the second router is configured as an egress node. An egress node

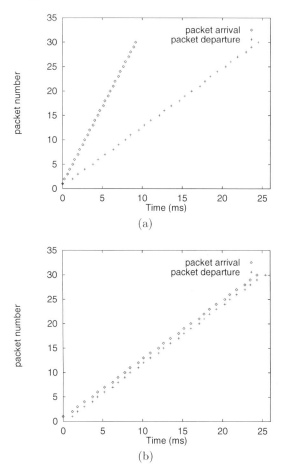

Fig. 5.8. Packet arrival and departure times for a 10 Mbps flow at (a) the ingress node, and (b) the egress node.

also implements the functionalities of a core node. In addition, it restores the initial values of the *ip_off* field. All traffic is UDP and all packets are 1000 bytes, not including the header.

In the first experiment we consider a flow between hosts 1 and 3 that has a reservation of 10 Mbps but sends at a much higher rate of about 30Mbps. Figures 5.8(a) and (b) plot the arrival and departure times for the first 30 packets of the flow at the ingress and egress node, respectively. One thing to notice in Figure 5.8(a) is that the arrival rate at the ingress node is almost three times the departure rate, which is the same as the reserved rate of 10 Mbps. This illustrate the non work conserving nature of the CJVC algorithm, which enforces the traffic profile and allows only 10 Mbps traffic into the network. Another thing to notice is that all packets incur about 0.8 ms delay in the egress node. This is because they are sent by the ingress node as soon as they become eligible, and therefore $g \simeq l/r = 8 \times 1052$ bits/10Mbps $= 0.84$ ms. As a result, they will be held in the rate-controller for this amount of time at the next hop[3], which is the egress node in our case.

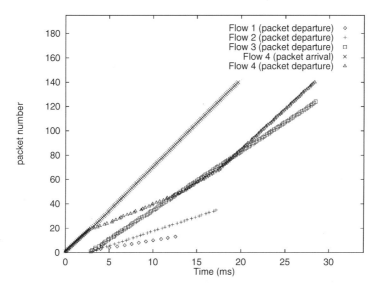

Fig. 5.9. The packets' arrival and departure times for four flows. The first three flows are guaranteed, with reservations of 10 Mbps, 20 Mbps, and 40 Mbps. The last flow is best effort with an arrival rate of about 60 Mbps.

In the second experiment we consider three guaranteed flows between hosts 1 and 3 with reservations of 10 Mbps, 20 Mbps, and 40 Mbps, respectively. In addition, we consider a fourth UDP flow between hosts 2 and 4 which is treated as best effort. The arrival rates of the first three flows are slightly larger than their reservations, while the arrival rate of the fourth flow is approximately 60 Mbps. At time 0, only the best-effort flow is active. At time 2.8 ms, the first three flows become simultaneously active. Flows 1 and 2 terminate after sending 12 and 35 packets, respectively. Figure 5.9 shows the packet arrival and departure times for the best-effort flow 4, and the packet departure times for the real-time flows 1, 2, and 3. As can be seen, the best-effort packets experience very low delay in the initial period of 2.8 ms. After the guaranteed flows become active, best-effort packets experience longer delays while guaranteed flows receive service at their reserved rate. After flow 1 and 2 terminate, the best-effort traffic grabs the remaining bandwidth.

The last two experiments illustrate the algorithms for admission control described in Section 5.4.3. The first experiment demonstrates the accuracy

---

[3]Note that since all packets have the same size, $\delta = 0$.

of estimating the aggregate reservation based on the b values carried in the packet headers. The second experiment illustrates the computation of the aggregate reservation bound, R_{bound}, when a new reservation is accepted or a reservation terminates. In these experiments we use an averaging interval, T_W, of 5 seconds, and a maximum inter-departure time, T_I, of 500 ms. Because all packets have the same size, the ingress to egress delays experienced by any two packets of the same flow are practically the same. As a result, we neglect the delay jitter, i.e., we assume $T_J = 0$. This gives us $f = (T_I + T_J)/T_W = 0.1$.

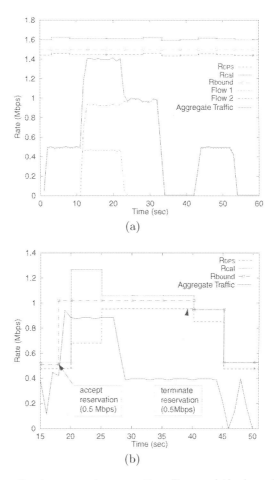

Fig. 5.10. The estimate aggregate reservation R_{cal}, and the bounds R_{bound} and R_{cal} in the case of (a) two ON-OFF flows with reservations of 0.5 Mbps, and 1.5 Mbps, respectively, and in the case when (b) one reservation of 0.5 Mbps is accepted at time $t = 18$ seconds, and then is terminated at $t = 39$ seconds.

In the first experiment we consider two flows, one with a reservation of 0.5 Mbps, and the other with a reservation of 1.5 Mbps. Figure 5.10(a) plots the arrival rate of each flow, as well as the arrival rate of the aggregate traffic. In addition, Figure 5.10(a) plots the bound of the aggregate reservation used by admission test, R_{bound}, the estimate of the aggregate reservation R_{DPS}, and the bound R_{cal} used to recalibrate R_{bound}. According to the pseudocode in Figure 5.6, both R_{DPS} and R_{cal} are updated at the end of each esti- mation interval. More precisely, every 5 seconds R_{DPS} is computed based on the b values carried in the packet headers, while R_{cal} is computed as $R_{DPS}/(1-f) + R_{new}$. Note that since, in this case, no new reservation is ac- cepted, we have $R_{new} = 0$, which yields $R_{cal} = R_{DPS}/(1-f)$. The important thing to note in Figure 5.10(a) is that the rate variation of the actual traffic (represented by the continuous line) has little effect on the accuracy of com- puting the aggregate reservation estimate R_{DPS}, and consequently of R_{cal}. In contrast, traditional measurement based admission control algorithms, which base their estimation on the *actual* traffic, would significantly underestimate the aggregate reservation, especially during the time periods when no data packets are received. In addition, note that since in this experiment R_{cal} is always larger than R_{bound}, and no new reservations are accepted, the value of R_{bound} is never updated.

In the second experiment we consider a scenario in which a new reservation of 0.5 Mbps is accepted at time $t = 18$ sec and terminates approximately at time $t = 39$ sec. For the entire time duration, plotted in Figure 5.10(b), we have background traffic with an aggregate reservation of 0.5 Mbps. Similar to the previous case, we plot the rate of the aggregate traffic, and, in addition, R_{bound}, R_{cal}, and R_{DPS}. There are several points worth noting. First, when the reservation is accepted at time $t = 18$ sec, R_{bound} increases by the value of the accepted reservation, i.e., 0.5 Mbps (see Figure 5.6). In this way, R_{bound} is guaranteed to remain an upper bound of the aggregate reservation R. In contrast, since both R_{DPS} and R_{cal} are updated only at the end of the estimation interval, they underestimate the aggregate reservation, as well as the aggregate traffic, before time $t = 20$ sec. Second, after R_{cal} is updated at time $t = 20$ sec, as $R_{DPS}/(1-f) + R_{new}$, the new value significantly overestimates the aggregate reservation. This is the main reason for which we do not use R_{cal} ($+R_{new}$), but R_{bound}, to do the admission control test. Third, note that unlike the case when the reservation was accepted, R_{bound} does not change when the reservation terminates at time $t = 39$ sec. This is simply because in our implementation no tear-down message is generated when a reservation terminates. However, as R_{cal} is updated at the end of the next estimation interval (i.e., at time $t = 45$ sec), R_{bound} drops to the correct value of 0.5 Mbps. This shows the importance of using R_{cal} to recalibrate R_{bound}. In addition, this illustrates the robustness of our algorithm, i.e., the overestimation in a previous period is corrected in the next period. Finally, note that in both experiments R_{DPS} always underestimates the aggregate reservation. This is due to the truncation errors in computing both the b values and the R_{DPS} estimate.

5.5.1 Processing Overhead

	Baseline		1 flow				10 flows				100 flows			
			ingress		egress		ingress		egress		ingress		egress	
	avg	std	avg	std	avg	std	avg	std	avg	std	avg	std	avg	std
enqueue	1.03	0.91	5.02	1.63	4.38	1.55	5.36	1.75	4.60	1.60	5.91	1.81	5.40	2.33
dequeue	1.52	1.91	3.14	3.27	2.69	2.81	2.79	3.68	2.30	2.91	2.77	2.82	1.73	2.12

Table 5.4. The average and standard deviation of the enqueue and dequeue times, measured in μs.

To evaluate the overhead of our algorithm we have performed three experiments on a 300 MHz Pentium II involving 1, 10, and 100 flows, respectively. The reservation and actual sending rates of all flows are identical. The aggregate sending rate is about 20% larger than the aggregate reservation rate. Table 5.4 shows the means and the standard deviations for the enqueue and dequeue times at both ingress and egress nodes. Each of these numbers is based on a measurement of 1000 packets. For comparison we also show the enqueue and dequeue times for the unmodified code. There are several points worth noting. First, our implementation adds less than 5 μs overhead per enqueue operation, and about 2 μs per dequeue operation. In addition, both the enqueue and dequeue times at the ingress node are greater than at the egress node. This is because ingress node performs per flow operations. Furthermore, as the number of flows increases, the enqueue times increase only slightly, i.e., by less than 20%. This suggests that our algorithm is indeed scalable in the number of flows. Finally, the dequeue times actually *decrease* as the number of flows increases. This is because the rate-controller is implemented as a calendar queue with each entry corresponding to a 128 μs time interval. Packets with eligible times falling between the same interval are stored in the same entry. Therefore, when the number of flows is large, more packets are stored in the same calendar queue entry. Since all these packets are transferred during one operation when they become eligible, the actual overhead per packet decreases.

5.6 Related Work

The idea of implementing guaranteed services by using a stateless core architecture was proposed by Jacobson [76] and Clark [24], and is now being pursued by the IETF Diffserv working group [32]. There are several differences between our scheme and the existing Diffserv proposals. First, our DPS based algorithms operate at a much finer granularity both in terms of time and traffic aggregates: the state embedded in a packet can be highly dynamic, as it encodes the *current* state of the flow, rather than the static and global properties such as dropping or scheduling priority. In addition, the goal of our scheme is to implement distributed algorithms that try to approximate the services provided by a network in which all routers implement per flow management. Therefore, we can provide service differentiation and performance guarantees in terms of both delay and bandwidth on a *per flow* basis. In contrast, the Premium service can provide only per flow bandwidth guarantees. Finally, we propose fully distributed and dynamic algorithms for implementing both data and control functionalities, where existing Diffserv solutions rely on more centralized and static algorithms for implementing admission control.

In this chapter, we propose a technique to estimate the aggregate reservation rate and use that estimate to perform admission control. While this may look similar to measurement-based admission control algorithms [62, 110], the objectives and thus the techniques are quite different. The measurement-based admission control algorithms are designed to support controlled-load type of services, the estimation is based on the *actual* amount of traffic transmitted in the past, and is usually an *optimistic* estimate in the sense that the estimated aggregate rate is smaller than the aggregate reserved rate. While this has the benefit of increasing the network utilization by the controlled-load service traffic, it has the risk of incurring transient overloads that may cause the service degradation. In contrast, our algorithm aims to support guaranteed service, and the goal is to estimate a close upper bound on the aggregate *reserved* rate even when the the actual arrival rate may vary.

Cruz [29] proposed a novel scheduling algorithm called SCED+ in the context of ATM networks. In SCED+, virtual circuits sharing a same path segment are aggregated into a virtual path. At each switch, only per virtual path state instead of per virtual circuit state needs to be maintained for scheduling purpose. In addition, an algorithm is proposed to compute the eligible times and the deadlines of a packet at subsequent nodes when the packet enters a virtual path. We note that by doing this and using DPS to carry this information in the packets' headers, it is possible to remove per path scheduling state from core nodes. However, unlike our solution, SCED+ does not provide per flow delay differentiation within an aggregate. In addition, the SCED+ work focuses on the data path mechanism, while we addresses both data path and control path issues.

5.7 Summary

In this chapter, we have described two distributed algorithms that implement QoS scheduling and admission control in a SCORE network. Combined, these two algorithms significantly enhance the scalability of both the data and control planes, while providing guaranteed services with flexibility, utilization, and assurance levels similar to those traditionally implemented with per flow mechanisms. The key technique used in both algorithms is Dynamic Packet State (DPS), which provides lightweight and robust means for routers to coordinate actions and implement distributed algorithms. By presenting a design and prototype implementation of the proposed algorithms in IPv4 networks, we have demonstrated that it is indeed possible to apply DPS techniques and have minimum incompatibility with existing protocols.

6 Providing Relative Service Differentiation in SCORE

In this chapter we describe a third application of the DPS technique: implementing a large spatial granularity network service, called Location Independent Resource Accounting (LIRA), that provides relative service differentiation. Unlike traditional services, such as the Guaranteed service, that are defined on a per flow basis, large spatial granularity services are defined over a large number of destinations. A simple example would be to guarantee a user 10 Mbps bandwidth irrespective of where or when the user sends traffic.

With LIRA, each user is assigned a *rate* at which it receives resource tokens. For each LIRA packet, a user is charged a number of resource tokens, the amount depending on the congestion level along the packet's path. The goal of LIRA is to achieve both high resource utilization and very low loss rate. LIRA provides relative differentiation: a user which receives twice as many resource tokens as another user will receive about twice as much bandwidth, as long as both users share the same links. Note that in the case of one link, LIRA reduces to Weighted Fair Queueing, i.e., each active user is allocated a capacity that is proportional to the rate at which it receives resource tokens.

We present an integrated set of algorithms that implement the LIRA service model in a SCORE network. Specifically, we leverage the existing routing infrastructure to distribute the path costs to all edge nodes. Since the path cost reflects the congestion level along the path, we use this cost to design dynamic routing and load balancing algorithms. To avoid packet re-ordering within a flow, we devise a lightweight mechanism based on DPS that binds a flow to a route so that all packets from the flow will traverse the same route. To reduce route oscillation, we probabilistically bind a flow to one of the multiple routes.

Traditional solutions to bind a flow to a route, also known as route-pinning, require routers to either maintain per flow state, or maintain state that is proportional with the square of the number of edge routers. By using DPS, we are able to significantly reduce this complexity. In particular, we propose a route-pinning mechanism that requires routers to maintain state which is proportional only to the number of egress routers.

The rest of the chapter is organized as follows. The next section motivates the LIRA service model, and discusses the limitation of the existing

I. Stoica: Stateless Core, LNCS 2979, pp. 103-127, 2004.
© Springer-Verlag Berlin Heidelberg 2004

alternatives. Section 6.3 describes the LIRA service and outlines its implementation in a SCORE network. Section 6.4 presents simulation experiments to demonstrate the effectiveness of our solution. Section 6.5 justifies the new service model and discusses possible ways for our scheme to implement other differential service models. Finally, in Section 6.6 we present the related work, and in Section 6.7 we summarize our contributions.

6.1 Background

Traditional service models that propose to enhance the best-effort service are usually defined on a per-flow basis. Examples of such services are the Guaranteed and Controlled Load services [93, 121] proposed in the context of Intserv [82], and the Premium service [76] proposed in the context of Diffserv [32]. While these services provide excellent support for a plethora of new point-to-point applications such as IP telephony and remote diagnostics, there is a growing need to support services at a coarser granularity than at a flow granularity. An example would be a service that is defined irrespective of *where* or *when* a user sends its traffic. Such a service would be much easier to negotiate by an organization since there is no need to specify the destinations in advance, usually a daunting task in practice.[1]

An example of such a service is the Assured service which was recently proposed by Clark and Wroclawski [23, 24] in the context of Diffserv. With the Assured service, a *fixed* bandwidth profile is associated with each user. This profile describes the commitment of the Internet Service Provider (ISP) to the user. In particular, the user is guaranteed that as long as its aggregate traffic does not exceed its profile, all user's packets are delivered to their destinations with very high probability. In the remainder of this chapter, we use the term of *service assurance* to denote the probability with which a packet is delivered to its destination. If a user exceeds its profile, the excess traffic is forwarded as best-effort traffic. Note that the implicit assumption in the Assured service is that the the traffic sent within a user profile has a much higher assurance (i.e., its packets are delivered with a much higher probability) than the best effort traffic.

In this chapter, we propose a novel service, called *Location Independent Resource Accounting* (LIRA), in which the service profile is described in terms of resource tokens rather than fixed bandwidth profile. In particular, with LIRA, each user is assigned a token bucket in which it receives resource tokens at a fixed rate. When a user sends a packet into the network, the user is charged a number of resource tokens, the amount depending on the congestion level along the path traversed by the packet. If the user does not have enough resource tokens in its token bucket, the packet is forwarded as

---

[1]For example, for a Web content provider, it is very hard if not impossible to specify its clients a priori.

a best effort packet. Note that, unlike the Assured service which provides an absolute service, LIRA provides a *relative* service. In particular, if a user receives resource tokens at a rate that is twice the rate of another user, and if both users sent traffic along the same paths, the first user will get twice as much aggregate throughput.

A natural question is why use a service that offers only relative bandwidth differentiation such as LIRA, instead of a service that offers a fixed bandwidth profile such as the Assured service? After all, the Assured service arguably provides a more powerful and useful abstraction; ideally, a user is guaranteed a fixed bandwidth irrespective of where or when it sends traffic. In contrast, with LIRA, the amount of traffic a user can send varies as a result of the congestion along the paths to the destination.

The simple answer is that, while the fixed bandwidth profile is arguably more powerful, it is unclear whether it can be efficiently implemented. The main problem follows directly from the service definition, as a fixed bandwidth profile service does not put any restriction on where or when a user can send traffic. This results in a fundamental conflict between maximizing resource utilization and achieving a high service assurance. In particular, since the network does not know in advance where the packets will go, in order to provide high service assurance, it needs to provision enough resources to *all possible* destinations. In the worst case, when the traffic of all users traverses the same congested link in the network, an ISP has to make sure that the sum of all user profiles does not exceed the capacity of the bottleneck link. Unfortunately, this will result in severe resource underutilization, which is unacceptable in practice. Alternatively, an ISP can provision resources for the average rather than the worst case scenario. Such a strategy will increase the resource utilization at the expense of service assurance.

In contrast, LIRA can achieve both high service assurance and resource utilization. However, to achieve this goal, LIRA gives up the fixed bandwidth profile semantics. The bandwidth profile of a user depends on the congestion in the network: the more congested the network, the lower the profile of a user. Thus, while the Assured service trades the service assurance for resource utilization, LIRA trades the fixed bandwidth service semantics for resource utilization. Next, we argue that the trade-off made by LIRA gives a user better control on managing its traffic, which makes LIRA a compelling alternative to the Assured service.

To illustrate this point, consider the case when the network becomes congested. In this case, LIRA tries to maintain the level of service assurance by scaling back the profiles of all users that send traffic in the congested portion of the network. In contrast, in the case of the Assured service, network congestion will cause a decrease of the service assurance for all users that share the congested portion of the network. Consider now a company whose traffic profile decreases from 10 to 5 Mbps, as a result of network congestion. Similarly, assume that, in the case of the Assured service, the same company experiences a ten fold increase in its loss rate as the result of the network congestion (while its service profile remains constant at 10 Mbps). Finally, assume that the CEO of the company wants to make an urgent video conference call, for which requires 2 Mbps. With LIRA, since the bandwidth required by the video conference is no larger than the company's traffic profile, the CEO can initiate the conference immediately. In contrast, with the Assured service, the CEO may not be able to start the conference due to the high loss rate. Worse yet, if the congestion is caused by the traffic of other users, the company can do nothing about it. The fundamental problem is that, unlike LIRA, the Assured service does not provide any protection in case of congestion.

6.2 Solution Outline

We consider the implementation of LIRA in a SCORE network, in which we use a two bit encoding scheme. The first bit, called the *preferred* bit, is set by the application or user and indicates the dropping preference of the packet. The second bit, called *marking* bit, is set by the ingress routers of an ISP and indicates whether the packet is in- or out-of-profile. When a preferred packet arrives at an ingress node, the node marks it, if the user has not exceeded its profile; otherwise the packet is left unmarked.[2] The reason to use two bits instead of one is that in an Internet environment with multiple ISPs, even if a packet may be out-of-profile in some ISPs on the earlier portion of its path, it may still be in-profile in a subsequent ISP. Having a dropping bit that is unchanged by upstream ISPs on the path will allow downstream ISPs to make the correct decision. Core routers implement a simple behavior of priority-based dropping. Whenever there is a congestion, a core router always drops unmarked packets first. In this chapter, we focus on mechanisms for implementing LIRA in a single ISP. We assume the following model for the interaction of multiple ISPs: if ISP A is using the service of ISP B, then ISP B will treat ISP A just like a regular user. In particular, the traffic from all ISP A's users will be treated as a single traffic aggregate.

While the above forwarding scheme can be easily implemented in a Diffserv network, it turns out that to effectively support LIRA we need the ability to perform route-pinning, that is, to bind a flow to a route so that all packets from the flow will traverse the same path. Unfortunately, traditional mechanisms to achieve route pinning require per flow state. Even the recently proposed Multi Protocol Label Switching (MPLS) requires routers to maintain an amount of state proportional to the square of the number of edge routers. In a large domain with thousands of edge nodes such overhead may be unacceptable.

---

[2]In this chapter, we will use the terminology of *marked* or *unmarked* packets to refer to packets in or out-of the service profile, respectively.

To address this problem we use the Dynamic Packet State (DPS) technique. With each packet we associate a label that encodes the packet's route from the current node to the egress router. Packet labels are initialized by ingress routers, and are used by core routers to route the packets. When a packet is forwarded, the router updates its label to reflect the fact that the remaining path has been reduced by one hop. By using this scheme, we are able to significantly reduce the state maintained by core routers. More precisely, this state becomes proportional to the number of egress nodes reachable from the core router, which can be shown to be optimal. The route pinning mechanism is described in detail in Section 6.3.4.

6.3 LIRA: Service Differentiation Based on Resource Right Tokens

In this section, we present our differential service model, called LIRA (Location Independent Resource Accounting), with service profiles defined in terms of resource tokens rather than absolute amounts of bandwidth.

With LIRA, each user i is assigned a service profile that is characterized by a resource token bucket (r_i, b_i), where r_i represents the resource token rate, and b_i represents the depth of the bucket. Unlike traditional token buckets where each preferred bit entering the network consumes exactly one token, with resource token buckets the number of tokens needed to admit a preferred bit is a dynamic function of the path it traverses.

Although there are many functions that can be used, we consider a simple case in which each link i is assigned a cost, denoted $c_i(t)$, which represents the amount of resource tokens charged for sending a marked bit along the link at time t. The cost of sending a marked packet is computed as $\sum_{i \in P} L \times c_i(t)$, where L is the packet length and P is the set of links traversed by the packet. While we focus on unicast communications in this chapter, we note that the cost function is also naturally applicable to the case of multicast. As we will show in Section 6.4, charging a user for every link it uses and using the cost in routing decisions helps to increase the network throughput. In fact, it has been shown by Ma et al. [68] that using a similar cost function[3] for performing the shortest path routing gives the best overall results when compared with other dynamic routing algorithms.

It is important to note that the *costs* used here are not monetary in nature. Instead they are reflecting the level of congestion and the resource usage along links/paths. This is different from a pricing scheme which represents the amount of payment made by an individual user. Though costs can provide

---

[3]It can be shown that when all links have the same capacity our cost is within a constant factor from the cost of shortest-dist(P, 1) algorithm proposed by Ma et al. [68].

valuable input to pricing policies, in general, there is *no* necessary direct connection between cost and price.

Figure 6.1 illustrates the algorithm performed by ingress nodes. When a preferred packet arrives at an ingress node, the node computes its cost based on the packet length and the path it traverses. If the user has enough resource tokens in its bucket to cover this cost, the packet is marked, admitted in the network, and the corresponding number of resource tokens is subtracted from the bucket account. Otherwise, depending on the policy, the packet can be either dropped, or treated as best effort. Informally, our goal at the user level is to ensure that users with "similar" communication patterns receive service (in terms of aggregate marked traffic) in proportion to their token rates.

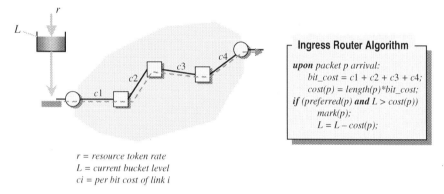

Ingress Router Algorithm

upon *packet p arrival:*
    $bit\_cost = c1 + c2 + c3 + c4;$
    $cost(p) = length(p)*bit\_cost;$
if *(preferred(p)* and $L > cost(p)$)
    $mark(p);$
    $L = L - cost(p);$

r = resource token rate
L = current bucket level
c_i = per bit cost of link i

Fig. 6.1. When a preferred packet arrives, the node computes the packet's cost, and the packet is marked if there are sufficient resource tokens.

The crux of the problem then is the computation and distribution of the per marked bit cost for each path. In this section, we first present the algorithm to compute the cost of each marked bit for a single *link*, and next present an algorithm that computes and distributes the per-path cost of one marked bit by leveraging existing routing protocols. We then argue that this dynamic cost information is also useful for multi-path routing and load balancing purposes. To avoid route oscillation and packet reordering within one application-level flow, we introduce two techniques. First, a lightweight scheme is devised to ensure that all packets from the same application-level flow always travel the same path. The scheme is lightweight in the sense that no per flow state is needed in any core routers. Second, rather than using a simple greedy algorithm that always selects the path with the current lowest cost, we use a probabilistic scheme to enhance system stability.

6.3.1 Link Cost Computation

A natural goal in designing the link cost function in LIRA is to avoid dropping marked packets. Since in the worst case all users can compete for the same link at the same time, a sufficient condition to avoid this problem is to have a cost function that exceeds the number of tokens in the system when the link utilization approaches unity. Without bounding the number of tokens in the system, this suggests a cost function that goes to infinity when the link utilization approaches unity. Among many possible cost functions that exhibit this property, we choose the following one:

$$c(t) = \frac{a}{1 - u(t)}, \qquad (6.1)$$

where a is the *fixed* cost of using the link[4] when it is idle, and $u(t)$ represents the link utilization at time t. In particular, $u(t) = R(t)/C$, were $R(t)$ is the traffic throughput at time t, and C represents the link capacity. Recall that $c(t)$ is measured in tokens/bit and represents how much a user is charged for sending a marked bit along that link at time t.

In an ideal system, where costs are instantaneously distributed and the rate of the incoming traffic varies slowly, a cost function as defined by Eq. (6.1) guarantees that no marked packets are dropped inside the core. However, in a real system, computing and distributing the cost information incur overhead, so they are usually done periodically. In addition, there is always the issue of propagation delay. Because of these, the cost information used in admitting packets at ingress nodes may be obsolete. This may cause packet dropping, and lead to oscillations. Though oscillations are inherent to any system in which the propagation of the feed-back information is non-zero, the sensitivity of our cost function when the link utilization approaches unity makes things worse. In this regime, an incrementally small traffic change may result in an arbitrary large cost change. In fact one may note that Eq. (6.1) is similar to the equation describing the delay behavior in queueing systems [65], which is known to lead to system instability when used as a congestion indication in a heavily loaded system.

To address these issues, we use the following iterative formula to compute the link cost:

$$c(t_i) = a + c(t_{i-1}) \frac{\widehat{R}(t_i, t_{i-1})}{C}. \qquad (6.2)$$

where $\widehat{R}(t', t'')$ denotes the average bit rate of the marked traffic during the time interval $[t', t'')$. It is easy to see that if the marked traffic rate is constant and equal to R, the above iteration converges to the cost given by Eq. (6.1).

---

[4]In practice, the network administrator can make use of a to encourage/discourage the use of the link. Simply by changing the fixed cost a, a link will cost proportionally more or less at the same utilization.

The main advantage of using Eq. (6.2) over Eq. (6.1) is that it is more robust against large variations in the link utilization. In particular, when the link utilization approaches unity the cost increases by at most a every iteration. In addition, unlike Eq. (6.1), Eq. (6.2) is well defined even when the link is congested, i.e., $\widehat{R}(t_{i-1}, t_i) = C$.

Unfortunately, computing the cost by using Eq. (6.2) is not as accurate as by using Eq. (6.1). The link may become and remain congested for a long time before the cost increase is large enough to reduce the arrival rate of marked bits. This may result in the loss of marked packets, which we try to avoid. To alleviate this problem we use only a fraction of the link capacity, $\bar{C} = \beta C$, for the marked traffic, the remaining being used to absorb the unexpected variations due to inaccuracies in the cost estimation.[5] Here, we chose β between 0.85 and 0.9.

6.3.2 Path Cost Computation and Distribution

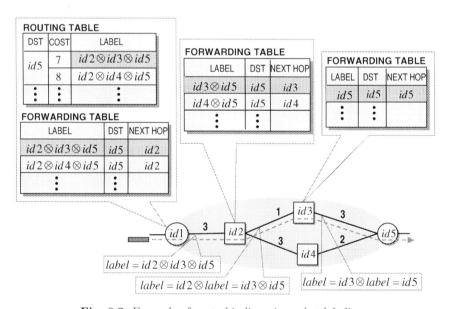

Fig. 6.2. Example of route binding via packet labeling.

In LIRA, the cost of a marked bit over a path is the sum of the costs of a marked bit over each link on the path. Once the cost for each link is computed, it is easy to compute and distribute the path cost by leveraging existing routing protocols. For link state algorithms, the cost of each marked

---

[5]β is similar to the pressure factor used in some ABR congestion control schemes for estimating the fair share [61, 87].

bit can be included as part of the link state. For distance vector algorithms, we can pass and compute the partial path cost in the same way the distance of a partial path is computed with respect to the routing metric.

6.3.3 Multipath Routing and Load Balancing

Since our algorithm defines a dynamic cost function that reflects the congestion level of each link, it is natural to use this cost function for the purpose of multi-path routing. To achieve this, we compute the k shortest paths for each destination or egress node using the unit link metric. While the obvious solution is to send packets along the path with the minimum cost (in the sense of LIRA, see Section 6.3) among the k paths, this may introduce two problems: (a) packet re-ordering within one application-level flow, which may negatively affect end-to-end congestion control algorithms, and (b) route oscillation, which may lead to system instability.

We introduce two techniques to address these problems. First, we present a lightweight mechanism that binds a flow to a route so that all packets from the flow will traverse the same route. Second, to reduce route oscillation, for each new flow, an ingress node probabilistically binds it to one of the multiple routes. By carefully selecting the probability, we can achieve both stability and load-balancing.

6.3.4 Route Pinning

As discussed earlier, we will maintain multiple routes for each destination. However, we would like to ensure that all packets belonging to the same flow are forwarded along the same path. To implement this mechanism we use the Dynamic Packet State (DPS) technique.

The basic idea is to associate with each path a label computed as the XOR over the identifiers of all routers along the path, and then associate this label with each packet of a flow that goes along that path. Here we use the IP address as the identifier. More precisely, a path $P = (id_0, id_1, \ldots, id_n)$, where id_0 is the source and id_n is the destination, is encoded at the source (id_0) by $l_0 = id_1 \otimes id_2 \otimes \ldots \otimes id_n$. Similarly, the path from id_1 to id_n is encoded at id_1 by $l_1 = id_2 \otimes \ldots \otimes id_n$. A packet that travels along path P is labeled with l_0 as it is leaving id_0, and with l_1 as it is leaving d_1. By using XOR we can iteratively re-compute the label based on the packet's current label and the node identifier. As an example, consider a packet that is assigned label l_0 at node id_0. When the packet arrives at node id_1, the new label corresponding to the remaining of the path, $(id_1, \ldots, id_n)$, is computed as follows:

$$l_1 = id_1 \otimes l_0 = \tag{6.3}$$
$$id_1 \otimes (id_1 \otimes id_2 \otimes \ldots \otimes id_n) = id_2 \otimes \ldots \otimes id_n.$$

It is easy to see that this scheme guarantees that the packet will be forwarded exactly along the path P. Here, we implicitly assume that all alternate paths between two end-nodes have unique labels. Although theoretically there is a non-zero probability that two labels may collide, we believe that for practical purposes it can be neglected. One possible way to reduce the label collision probability would be to use a hash function to translate the IP addresses into labels. By using a good hash function, this will result in a more random distribution of router labels. Another possibility would be to explicitly label routers to reduce or even eliminate the collision probability. Note that this solution will require to maintain the mapping between router IP addresses and router labels, which can be difficult in practice. One last point worth noting is that even if two alternate paths have the same label, this will not jeopardize the correctness of our scheme: the worst thing that can happen is an alternate path to be ignored, which will only lead to a decrease in utilization.

Next we give some details of how this mechanism can be implemented by simply extending the information maintained by each router in the routing and forwarding tables. Besides the destination and the route cost, each entry in the routing table also contains the label associated with that path.

$$< dst, < cost^{(1)}, l^{(1)} >, \ldots < cost^{(k)}, l^{(k)}) >> \qquad (6.4)$$

Similarly, the forwarding table should contain an entry for each path:

$$< l^{(1)}, dst, next\_hop^{(1)} > \ldots < l^{(k)}, dst, next\_hop^{(k)} > \qquad (6.5)$$

In Figure 6.2 we give a simple example to illustrate this mechanism. Assume that nodes id_1 and id_5 are edge nodes, and there are two possible paths from id_1 to id_5 of costs 7, and 8, respectively. Now, assume a packet destined to id_5 arrives at id_1. First the ingress node id_1 searches the classifier table (not shown in the Figure) that maintains a list of all flows to see whether this is the first packet of a flow. If it is, the router uses the information in the routing table to probabilistically bind the flow to a path to id_5. At the same time it labels the packet with the encoding of the selected route. In our example, assume the path of cost 7, i.e., (id_1, id_2, id_3, id_5), is selected. If the arriving packet is not the first packet of the flow, the router automatically labels the packet with the encoding of the path to which the flow is bound. This can be simply achieved by keeping a copy of the label in the classifier table. Once the packet is labeled, the router checks the forwarding table for the next hop by matching the packet's label and its destination. In our case, this operation gives us id_2 as the next hop. When the packet arrives at node id_2 the router first computes a new label based on the current packet label and the router identifier: $label = id_2 \otimes label$. The new label is then used to lookup the forwarding table.

It is important to note that the above algorithm assumes per flow state only at ingress nodes. Inside the core, there is no per flow state. Moreover, the labels can speed-up the table lookup if used as hash keys.

6.3.5 Path Selection

While the above forwarding algorithm ensures that all packets belonging to the same flow traverse the same path, there is still the issue of how to select a path for a new flow. The biggest concern with any dynamic routing protocol based on congestion information is its stability. Frequent route changes may lead to oscillations.

To address this problem, we associate a probability with each route and use it in binding a new flow to that route. The goal in computing this probability is to equalize the costs along the alternate routes, if possible. For this we use a greedy algorithm. Every time the route costs are updated we split the set of routes in two equal sets, where all the routes in one set have costs larger than the routes in the second set. If there is an odd number of routes, we leave the median out. Then, we decrease the probability of every route in the first set, the one which contains the higher cost routes, and increase the probability of each route in the second set by a small constant δ. It can be shown that in a steady-state system, this algorithm converges to the desired solution, in which the difference between the costs of the two alternate paths is bounded by δ.

6.3.6 Scalability

As described so far, it is required that our scheme maintains k entries for each destination in both the forwarding table used by the forwarding engine and the routing table used by the routing protocol, where k is the maximum number of alternate paths. While this factor may not be significant if k is small, a more serious issue that potentially limits the scalability of the algorithm is that in its basic form it requires that an entry be maintained for each destination, where in reality, to achieve scalability, routers really maintain the longest-prefix of a group of destinations that share the same route [41]. Since our algorithm works in the context of one ISP, we can maintain an entry for each *egress node* instead of each destination. We believe this is sufficient as the number of egress nodes in an ISP is usually not large.

However, assume that the number of egress nodes in an ISP is very large so that significant address aggregation is needed. Then we need to also perform cost aggregation. To illustrate the problem consider the example in Figure 6.3. Assume the addresses of d_0 and d_1 are aggregated at an intermediate router r_1. Now the question is how much to charge a packet that enters at the ingress node r_0 and has the destination d_0. Since we do not keep state for the individual routes to d_0, and d_1 respectively, we need to aggregate the cost to these two destinations. In doing this, a natural goal would be to maintain the total charges the same as in a reference system that keeps per route state. Let $R(r_1, d_i)$ denote the average traffic rate from r_1 to d_i, $i = 1, 2$. Then, in the reference system that maintains per route state, the total charge per time unit for the aggregate traffic from r_1 to d_0 and d_1 is: $cost(r_1, d_0)R(r_1, d_0) +$

$cost(r_1, d_1)R(r_1, d_1)$. In a system that does not maintain per route state, the charge for the same traffic is $cost(r_1, d_0, d_1)(R(r_1, d_0) + R(r_1, d_1))$, where $cost(r_1, d_0, d_1)$ denotes the per bit aggregate cost. This yields

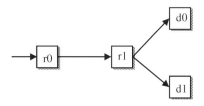

Fig. 6.3. Topology to illustrate the label and cost aggregation.

$$cost(r_1, d_0, d_1) = \frac{cost(r_1, d_0)R(r_1, d_0)}{R(r_1, d_0) + R(r_1, d_1)} + \qquad (6.6)$$
$$\frac{cost(r_1, d_1)R(r_1, d_1)}{R(r_1, d_0) + R(r_1, d_1)}.$$

Thus, any packet that arrives at r_0 and has either destination d_0 or d_1 is charged with $cost(r_0, r_1) + cost(r_1, d_0, d_1)$. Obviously, route aggregation increases the inaccuracies in cost estimation. However, this may be alleviated by the fact that the route aggregation usually exhibits high localities.

Another problem with address aggregation is that a label can no longer be used to encode the entire path to the destination. Instead, it is used to encode the *common* portion of the paths to the destinations in the aggregate set. This means that a packet should be relabeled at every router that performs aggregation involving the packet's destination. The most serious concern with this scheme is that it is necessary to maintain per flow state and perform packet classification at a core router (r_1 in our example). Fortunately, this scalability problem is alleviated by the fact that we need to keep per flow state *only* for the flows whose destination addresses are aggregated at the *current* router. Finally, we note that this problem is not specific to our scheme; any scheme that (i) allows multiple path routing, (ii) performs load balancing, and (iii) avoids packet reordering has to address it.

6.4 Simulation Results

In this section we evaluate our model by simulation. We conduct four experiments: three involving simple topologies which help to gain a better understanding of the behavior of our algorithms, and one more realistic example with a larger topology and more complex traffic patterns. The first experiment shows that if all users share the same congested path, then each user receives service in proportion to its resource token rate. This is the same result one would expect from using a weighted fair queueing scheduler on every link, with the weights set to the users' token rate. In the second experiment, we show that by using dynamic routing and load balancing, we are able to achieve the same result – that is, each user receives service in proportion to its token rate – in a more general configuration where simply using weighted fair queueing scheduler on every link is not sufficient. In the third experiment, we show how load balancing can significantly increase the overall resource utilization. Finally, the fourth experiment shows how the behaviors observed in the previous experiments scale to a larger topology.

6.4.1 Experiment Design

We have implemented a packet level simulator which supports both Distance Vector (DV) and Shortest Path First (SPF) routing algorithms. To support load balancing we extended these algorithms to compute the k-th shortest paths. The time interval between two route updates is uniformly distributed between 0.5 and 1.5 of the average value. As shown by Floyd and Jacobson [38], this choice avoids the route-update self-synchronization. In SPF, when a node receives a routing message, it first updates its routing table and then forwards the message to all its neighbors, except the sender. The routing messages are assumed to have high priority, so they are never lost. In the next sections we compare the following schemes:

- BASE – this scheme models today's best-effort Internet, and it is used as a baseline in our comparison. The routing protocol uses the number of hops as the distance metric and it is implemented by either DV or SPF. This scheme does not implement service differentiation, i.e., both marked and unmarked packets are identically treated.
- STATIC – this scheme implements the same static routing as BASE. In addition, it implements *LIRA* by computing the link cost as described in Section 6.3.1, and marking packets at each ingress node according to the algorithm shown in Figure 6.1.
- DYNAMIC-k – this scheme adds dynamic routing and load balancing to STATIC. The routing protocol uses a modified version of DV/SPF to find the first k shortest paths. Note that DYNAMIC-1 is equivalent to STATIC.

Each router implements a FIFO scheduling discipline with a shared buffer and a drop-tail management scheme. When the buffer occupancy exceeds a predefined threshold, newly arrived unmarked packets are dropped. Thus, the entire buffer space from the threshold up to its total size is reserved to the in-profile traffic.[6] Unless otherwise specified, throughout all our experiments we use a buffer size of 256 KB and a threshold of 64 KB.

---

[6]We note that this scheme is a simplified version of the RIO buffer management scheme proposed by Clark and Wroclawski [24] In addition, RIO implements a Random Early Detection (RED) [37] dropping policy, instead of drop-tail, for both

The two main performance indices that we use in comparing the above schemes are the user *in-profile* and user *overall* throughputs. The user in-profile throughput represents the rate of the user aggregate *in-profile* traffic delivered to its destinations. The overall throughput represents the user's *entire* traffic — i.e., including both the in- and out-of profile traffic — delivered to its destinations. In addition, we use user *dropping rate* of the in-profile traffic to characterize the level of service assurance.

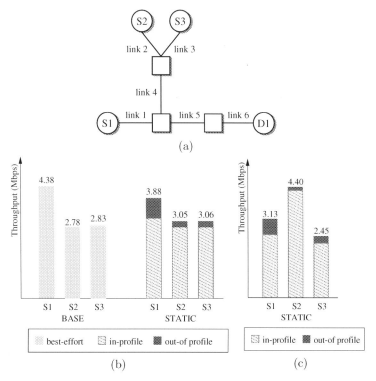

Fig. 6.4. (a) Topology used in the first experiment. Each link has 10 Mbps capacity. $S1$, $S2$, and $S3$ send all their traffic to $D1$. (b) The throughputs of the three users under BASE and STATIC schemes. (c) The throughputs under STATIC when the token rate of $S2$ is twice the rate of $S1/S2$.

Recent studies have shown that the traffic in real networks exhibits the self-similar property [27, 80, 81, 118] — that is, the traffic is bursty over widely different time scales. To generate self-similar traffic we use the tech-

---

in- and out-of profile traffic. RED provides an efficient detection mechanism for the adaptive flows, such as TCP, allowing them to gracefully degrade their performances when congestion occurs. However, since in this study we are not concerned with the behavior of individual flows, for simplicity we chose to not implement RED.

nique originally proposed by Willinger et al. [118], where it was shown that the superposition of many ON-OFF flows with ON and OFF periods drawn from a heavy tail distribution, and which have fixed rates during the ON period results in self-similar traffic. In particular, Willinger et al. [118] show that the aggregation of several hundred of ON-OFF flows is a reasonable approximation of the real end-to-end traffic observed in a LAN.

In all our experiments, we generate the traffic by drawing the length of the ON and OFF periods from a Pareto distribution with the power factor of 1.2. During the ON period a source sends packets with sizes between 100 and 1000 bytes. The time to send a packet of minimum size during the ON period is assumed to be the time unit in computing the length of the ON and OFF intervals.

Due to the high overhead incurred by a packet-level simulator such as ours, we limit the link capacities to 10 Mbps and the simulation time to 200 sec. We set the average interval between routing updates to 5 sec for the small topologies used in the first three experiments, and to 3 sec for the large topology used in the last experiment. In all experiments, the traffic starts at time $t = 20$ sec. The choice of this time guarantees that the routing algorithm finds at least one path between any two nodes by time t. In order to eliminate the transient behavior, we start our measurements at time $t = 50$ sec.

6.4.2 Experiment 1: Local Fairness and Service Differentiation

This experiment shows that if all users send their traffic along the same congested path, they get service in proportion to their token rate, as long as there is enough demand. Consider the topology in Figure 6.4(a), where users $S1$, $S2$, and $S3$ send traffic to $D1$. Figure 6.4(b) shows the user overall throughputs over the entire simulation under BASE. As it can be seen, $S1$ gets significantly more than the other two. In fact, if the traffic from all sources were continuously backlogged, we expect that $S1$ will get half of the congested links 5 and 6, while $S2$ and $S3$ split the other half. This is because even though each user sends at an average rate higher than 10 Mbs, the queues are not continuously backlogged. This is due to the bursty nature of the traffic and due to the limited buffer space at each router.

Next, we run the same simulation for the STATIC scheme. To each user we assign the same token rate, and to each link we associate the same fixed cost. Figure 6.4(b) shows the user overall and in-profile throughputs. Compared to BASE, the overall throughputs are more evenly distributed. However, the user $S1$ still gets slightly better service, i.e., its in-profile throughput is 3.12 Mbps, while the in-profile throughput of $S2/S3$ is 2.75 Mbps. To see why, recall from Eq. (6.1) that link cost accurately reflects the level of congestion on that link. Consequently, in this case links 5 and 6 will have the highest cost, followed by link 4, and then the other three links. Thus, $S2$ and $S3$ have to "pay" more than $S1$ per marked bit. Since all users have the same

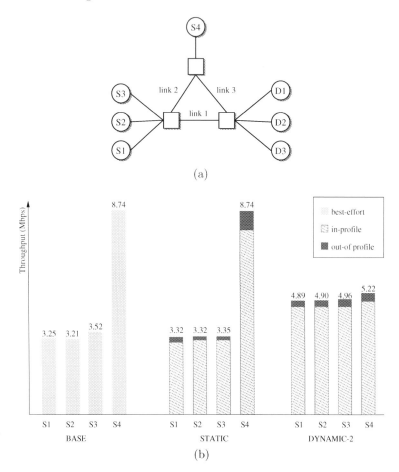

(a)

(b)

Fig. 6.5. (a) Topology used in the second experiment. $S1$, $S2$, $S3$, and $S4$ send all their traffic to $D1$, $D2$, and $D3$, respectively. (b) The throughputs of all users under BASE, STATIC, and DYNAMIC-2.

token rates, this translates into lower overall throughputs for $S2$ and $S3$, respectively.

To illustrate the relationship between the user's token rate and its performance, we double the token rate of $S2$. Figure 6.4(c) shows the overall and in-profile throughputs of each user. In terms of in-profile traffic, user $S2$ gets roughly twice the throughout of $S3$ (i.e., 4.27 Mbps vs. 2.18 Mbps).

Finally, we note that there were no marked packets dropped in any of the above simulations. For comparison, more than 60% of the out-of profile traffic was dropped.

6.4.3 Experiment 2: User Fairness and Load Balancing

In this section we show how dynamic routing and load balancing help to improve user level fairness and achieve better resource utilization. Consider the topology in Figure 6.5 where users $S1$, $S2$, $S3$ and $S4$ send traffic to each of the users $D1$, $D2$ and $D3$, respectively. Again the fixed costs of all links are equal, and all users are assigned the same token rate.

Figure 6.5(b) shows the overall and in-profile throughputs of $S1$, $S2$, $S3$ and $S4$ under BASE, STATIC and DYNAMIC-2, respectively. When BASE and STATIC are used, each user always sends along the shortest paths. This results in $S1$, $S2$ and $S3$ sharing link 1, while $S4$ alone uses link 3. As a consequence $S4$ receives significantly better service than the other three users. Since it implements the same routing algorithm, STATIC does not improve the overall throughputs. However, compared with BASE, STATIC guarantees that in-profile packets are delivered with very high probability (again, in this experiment, no marked packets were dropped). On the other hand, when DYNAMIC-2 is used, each user receives almost the same service. This is because users $S1$, $S2$ and $S3$ can now use both routes to send their traffic, which allows them to compete with user $S4$ for link 3. User $S4$ still maintains a slight advantage, but now the difference between its overall throughput and the overall throughputs of the other users is less than 7%. In the case of the in-profile traffic this difference is about 5%. As in the previous experiment, the reason for this difference is because when competing with $S4$, the other users have to pay, besides link 3, for link 2 as well.

Thus, by taking advantage of the alternate routes, our scheme is able to achieve fairness in a more general setting. At the same time it is worth noting that the overall throughput also increases by almost 7%. However, in this case, this is mainly due to the bursty nature of $S4$'s traffic which cannot use the entire capacity of link 3 when it is the only one using it, rather than load balancing.

6.4.4 Experiment 3: Load Distribution and Load Balancing

This experiment shows how the load distribution affects the effectiveness of our load balancing scheme. For this purpose, consider the topology in Figure 6.6(a). In the first simulation we generate flows that have the source and the destination uniformly distributed among users. Figure 6.6(b) shows the means of the overall throughputs under BASE, STATIC, and DYNAMIC-2, respectively.[7] Due to the uniformity of the traffic pattern, in this case BASE performs very well. Under STATIC we get slightly larger overall throughput, mainly due to our congestion control scheme, which admits a marked packet

---

[7]We have also computed standard deviations for each case: the largest standard deviation was 0.342 for the overall throughput under STATIC scheme, and 0.4 for the in-profile throughput under DYNAMIC-2.

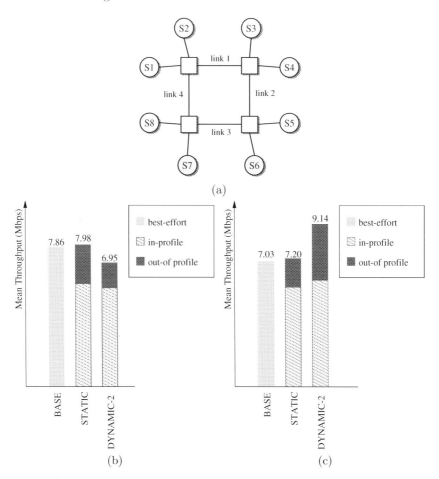

Fig. 6.6. (a) Topology used in the third experiment. Mean throughputs when (b) load is balanced, and (c) when it is unbalanced, i.e, S3 and S4 are inactive.

only if there is a high probability that it will be delivered. However, under DYNAMIC-2 the performance degrades. This is because there are times when our probabilistic routing algorithm selects longer routes, which leads to inefficient resource utilization.

Next, we consider an unbalanced load by making users $S3$ and $S4$ inactive. Figure 6.6(c) shows throughput means under BASE, STATIC, and DYNAMIC-2, respectively. As it can be noticed, using DYNAMIC-2 increases the mean by 30%. This is because under BASE and STATIC schemes the entire traffic between $S1$, $S2$ and $S5$, $S6$ is routed through links 3 and 4 only. On the other hand, DYNAMIC-2 takes advantage of the alternate route through links 1 and 2.

Finally, in another simulation not shown here we considered the scenario in which $S5$, $S6$, $S7$, and $S8$ send their entire traffic to $S3$ and $S4$, respectively. In this case DYNAMIC-2 outperforms STATIC and BASE by almost two times in terms of in-profile and overall throughputs. This is again because BASE and STATIC exclusively use links 3 and 2, while DYNAMIC-2 is able to use all four links.

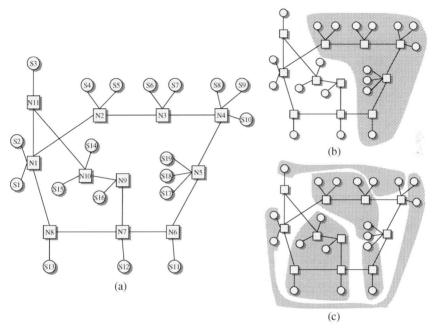

Fig. 6.7. Topology similar to the T3 topology of the NSFNET backbone network containing the IBM NSS nodes.

6.4.5 Experiment 4: Large Scale Example

In this section we consider a larger topology that closely resembles the T3 topology of the NSFNET backbone containing the IBM NSS nodes (see Figure 6.7). The major difference is that in order to limit the simulation time we assume 10 Mbps links, instead of 45 Mbps. We consider the following three scenarios.

In the first scenario we assume that load is uniformly distributed, i.e., any two users communicate with the same probability. Figure 6.8(a) shows the results for each scheme which are consistent with the ones obtained in the previous experiment. Due to the congestion control which reduces the number of dropped packets in the network, STATIC achieves higher throughput

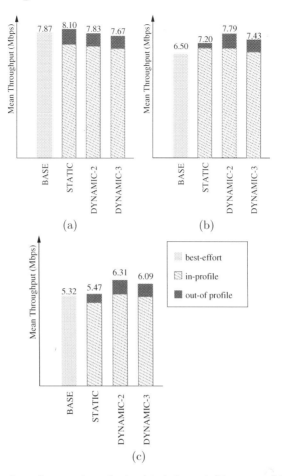

Fig. 6.8. The throughputs when the load is balanced (Figure 6.7(a)), (b) unbalanced ((Figure 6.7(b)), and (c) when the network is virtually partitioned (Figure 6.7(c)).

than BASE. On the other hand, dynamic routing and load balancing are not effective in this case, since they tend to generate longer routes which leads to inefficient resource utilization. This is illustrated by the decrease of the overall and the in-profile throughputs under DYNAMIC-2 and DYNAMIC-3, respectively.

In the second scenario we assume unbalanced load. More precisely, we consider 11 users (covered by the shaded area in Figure 6.7(b)) which are nine times more active than the others, i.e., they send/receive nine times more traffic.[8] Unlike the previous scenario, in terms of overall throughput, DYNAMIC-2 outperforms STATIC by almost 8%, and BASE by almost 20% (see Figure 6.8(b)). This is because DYNAMIC-2 is able to use some of the idle links from the un-shaded partition. However, as shown by the results for DYNAMIC-3, as the number of alternate paths increases both the overall and in-profile throughputs start to decrease.

In the final scenario we consider the partition of the network shown in Figure 6.7(c). For simplicity, we assume that only users in the same partition communicate between them. This scenario models a virtual private network (VPN) setting, where each partition corresponds to a VPN. Again, DYNAMIC-2 performs best[9] since it is able to make use of some links between partitions that otherwise would remain idle.

Finally, we note that across all simulations presented in this section, the dropping rate for the marked packets was never larger than 0.3%. At the same time the dropping rate for the unmarked packets was over 40%.

6.4.6 Summary of Simulation Results

Although the experiments in this section are far from being exhaustive, we believe that they give a reasonable image of how our scheme performs. First, our scheme is effective in providing service differentiation at the user level. Specifically, the first two experiments show that users with similar communication patterns get service in proportion to their token rates. Second, at least for the topologies and the traffic model considered in these experiments, our scheme ensures that marked packets are delivered to the destination with high probability.

Consistent with other studies [68], these experiments show that performing dynamic routing and load balancing make little sense when the load is already balanced. In fact, using dynamic routing and load balancing can actually hurt, since, as noted above, this will generate longer routes which may result in inefficient resource utilization. However, when the load is unbalanced, using DYNAMIC-k can significantly increase the utilization and achieve a higher degree of fairness.

Finally, we note that the in-profile dropping rate decreases as the the number of alternate paths increases. For example in the last experiment in the first two scenarios the dropping rate is no larger than 0.3% under STATIC and 0% under DYNAMIC-2 and DYNAMIC-3, respectively, while in the last scenario the percentage decreases from 0.129% for STATIC, to 0.101% for DYNAMIC-2, and to 0.054% for DYNAMIC-3.

---

[8]This might model the real situation where the east coast is more active than the west coast between 9 and 12 a.m. EST.

[9]The mean of the user overall throughput under DYNAMIC-2 is 15% larger than under STATIC, and 18% larger than under BASE.

6.5 Discussion

In this chapter, we have studied a differential service model, *LIRA*, in which, unlike the Assured service [23, 24], the service profile is specified in terms of resource tokens instead of absolute bandwidth. Since the exact bandwidth of marked bits that a customer can receive from such a service is not *known* a priori, a natural question to ask is why such a service model is interesting?

There are several reasons. First, we believe that the apriori specification of an absolute amount of bandwidth in the service profile, though desirable, is not essential. In particular, we believe that the essential aspects that distinguish Diffserv from Intserv are the following: (a) the service profile is used for traffic aggregates which are much coarser than per flow traffic, and (b) the service profile is defined over a timescale larger than the duration of individual flows, i.e. service profile is rather *static*. Notice that the degree of traffic aggregation directly relates to the spatial granularity of the service profile. On the one hand, if each service profile is defined for only one destination, we have the smallest degree of traffic aggregation. If there are N possible egress nodes for a user, N independent service profiles need to be defined. Network provisioning is relatively easy as the entire traffic matrix between all egress and ingress nodes is known. However, if a user has a rather dynamic distribution of egress nodes for its traffic, i.e., the amount of traffic destined to each egress node varies significantly, and the number of possible egress nodes is large, such a scheme will significantly reduce the chance of statistical sharing. On the other hand, if each service profile is defined for all egress nodes, we have the largest degree of traffic aggregation. Only one service profile is needed for each user regardless of the number of possible egress nodes. In addition to a smaller number of service profiles, such a service model also allows all the traffic from the same user, regardless of its destination, to *statistically* share the same service profile. The flip side is that it makes it difficult to provision network resources. Since the traffic matrix is not known apriori, the best-case scenario is when the network traffic is evenly distributed, and the worst-case scenario is when all traffic goes to the same egress router.

Therefore, it is very difficult, if not impossible, to design service profiles that (1) are static, (2) support coarse spatial granularity, (3) are defined in terms of absolute bandwidth, and at the same time achieve (4) high service assurance and (5) high resource utilization. Since we feel that (1), (2), (4) and (5) are the most important for differential services, we decided to give up (3).

Fundamentally, we want a service profile that is static and path-independent. However, to achieve high utilization, we need to explicitly address the fact that congestion is a local and dynamic phenomenon. Our solution is to have two levels of differentiation: (a) the user or service-profile level differentiation, which is based on resource token arrival rate that is static and path independent, and (b) the packet level differentiation, which is a simple priority between marked and unmarked packets and weighted fair share among marked packets. By dynamically setting the cost of each marked bit as a function of the congestion level of the path it traverses, we set up the linkage between the static/path-independent and the dynamic/path-dependent components of the service model.

A second reason our service model may be acceptable is that users may care more about the *differential* aspect of the service than the guaranteed bandwidth. For example, if user A pays twice as much as user B, user A would expect to have roughly twice as much traffic delivered as user B during congestion if they share same congested links. This is exactly what we accomplish in *LIRA*.

A third reason a fixed-resource-token-rate-variable-bandwidth service profile may be acceptable is that the user traffic is usually bursty over multiple time-scales [27, 80, 118]. Thus, there is a fundamental mismatch between an absolute bandwidth profile and the bursty nature of the traffic.

We do recognize the fact that it is desirable for both the user and the ISP to understand the relationship between the user's resource token rate and its expected capacity. This can be achieved by measuring the rate of marked bits given a fixed token rate. Both the user and the ISP can perform this measurement. In fact, this suggests two possible scenarios in which *LIRA* can be used to provide a differential service with an expected capacity defined in terms of absolute bandwidth. In the first scenario, the service is not transparent. Initially, the ISP will provide the user with the following relationship

$$\text{expected\_capacity} = f(\text{token\_rate}, \text{traffic\_mix}) \qquad (6.7)$$

based on its own prior measurement. The user will measure the expected capacity and then make adjustments by asking for an increase or a decrease in its resource token rate. In the second scenario, the service is transparent. Both the initial setting and the subsequent adjustments of the service profile in terms of token rate will be made by the ISP only.

Therefore, one way of thinking about our scheme is that it provides a flexible and efficient framework for implementing a variety of Assured Services. In addition, the dynamic link cost information and the statistics of the resource token bucket history provide good feedback both for individual applications to perform runtime adaptation, and for the user or the ISP to do proper accounting and provisioning.

6.6 Related Work

The LIRA service is highly influenced by Clark and Wroclawski's Assured service proposal [23, 24]. The key difference is that we define service profiles in units of resource tokens rather than absolute bandwidth. In addition, we propose a resource accounting scheme and an integrated set of algorithms to implement our service model.

Another related proposal is the User-Share Differentiation (USD) [116] scheme, which does *not* assume absolute bandwidth profiles. In fact, with USD, a user is assigned a share rather than a token-bucket-based service profile. For *each* congested link in the network traversed by the user's traffic, the user shares the bandwidth with other users in proportion to its share. The service provided is equivalent to one in which *each* link in a network implements a weighted fair queueing scheduler where the weight is the user's share. With USD, there is little correlation between the share of a user and the aggregate throughput it will receive. For example, two users that are assigned the same share can see drastically different aggregate throughputs. A user that has traffic for many destinations (thus traverse many different paths) can potentially receive much higher aggregate throughput than a user that has traffic for only a few destinations.

Waldspurger and Weihl have proposed a framework for resource management based on lottery tickets [113, 114]. Each client is assigned a certain number of tickets which encapsulate its resource rights. The number of tickets a user receives is similar to the user's income rate in *LIRA*. This framework was shown to provide flexible management for various single resources, such as disk, memory and CPU. However, they do not give any algorithm(s) to coordinate ticket allocation among multiple resources.

To increase resource utilization, in this chapter we propose performing dynamic routing and load balancing among the best k shortest paths between source and destination. In this context, one of the first dynamic routing algorithms, which uses the link delay as metric, was the ARPANET shortest path first [71]. Unfortunately, the sensitivity of this metric when the link utilization approaches unity resulted to relatively poor performances. Various routing algorithms based on congestion control information were proposed elsewhere [43, 46]. The unique aspect of our algorithm is that it combines dynamic routing, congestion control and load balancing. We also alleviate the problem of system stability which plagued many of the previous dynamic routing algorithms by defining a more robust cost function and probabilistically binding a flow to a route. We also note that our link cost is similar to the one used by Ma et al. [68]. In particular, it can be shown that when all links have the same capacity, our link cost is within a constant factor of the cost of shortest-dist(P, 1) algorithm presented Ma et al. [68]. It is worth noting that shortest-dist(P, 1) performed the best among all the algorithms studied there.

6.7 Summary

In this chapter we have proposed an Assured service model in which the service-profile is defined in units of resource tokens rather than the absolute bandwidth, and an accounting scheme that dynamically determines the number of resource tokens charged for each in-profile packet. We have presented a set of algorithms that efficiently implement the service model. In particular, we introduced three techniques: (a) distributing path costs to all edge nodes by leveraging the existing routing infrastructure; (b) binding a flow to a route (route-pinning); (c) multi-path routing and probabilistic binding of flows to paths to achieve load balancing.

To implement route-pinning, which is arguably the most complex technique of the three, we have used DPS. By using DPS, we have been able to efficiently implement the LIRA service model in a SCORE network. We have presented simulation results to demonstrate the effectiveness of the approach. To the best of our knowledge, this is the first complete scheme that explicitly addresses the issue of large spatial granularities.

7 Making SCORE More Robust and Scalable

While SCORE/DPS based solutions are much more scalable and, in the case of fail-stop failures, more robust than their stateful counterparts, they are less scalable and robust than the stateless solutions. The scalability of the SCORE architecture suffers from the fact that the network core cannot transcend trust boundaries, such as the boundary between two competing Internet Service Providers (ISPs). As a result, the high-speed routers on these boundaries must be stateful edge routers. The lack of robustness is because the malfunctioning of a single edge or core router that inserts erroneous state in the packet headers could severely impact the performance of an entire SCORE network.

In this chapter, we discuss an extension to the SCORE architecture, called "verify-and-protect", that overcomes these limitations. We achieve scalability by pushing the complexity all the way to end-hosts, and therefore eliminate the distinction between core and edge routers. To address the trust and robustness issues, all routers statistically verify that the incoming packets are correctly marked, i.e., that they carry *consistent* state. This approach enables routers to discover and isolate misbehaving end-hosts and routers. While this approach requires routers to maintain state for each flow that is verified, in practice, this does not compromise the scalability of core routers as the amount of state maintained by these routers is very small. In practice, as discussed in Section 7.3, the number of flows that a router needs to verify simultaneously – flows for which the router has to maintain state – is on the order of tens. We illustrate the "verify-and-protect" approach in the context of Core-Stateless Fair Queueing (CSFQ), by developing tests to accurately identify misbehaving nodes, and present simulation results to demonstrate the effectiveness of this approach.

The remainder of this chapter is organized as follows. The next section describes the failure model assumed throughout this chapter. Section 7.2 presents the components of the "verify-and-protect" approach, while Section 7.3 describes the details of the flow verification algorithm in the case of CSFQ. Section 7.4 proposes a robust test to identify the misbehaving nodes. Finally, Section 7.5 presents simulation results, while Section 7.6 summarizes our findings.

I. Stoica: Stateless Core, LNCS 2979, pp. 129-151, 2004.
© Springer-Verlag Berlin Heidelberg 2004

7.1 Failure Model

In this chapter, we assume a partial failure model in which a router or end-host misbehaves by sending packets carrying inconsistent information. A packet is said to carry inconsistent information (or state), if this information does not correctly reflect the flow behavior. In particular, with CSFQ, a packet is said to carry inconsistent information if the difference between the estimated rate in its header and the actual flow rate exceeds some predefined threshold (see Section 7.3.2 for details). We use a range test, instead of an equality test, to account for the rate estimation inaccuracies due to the delay jitter and the probabilistic dropping scheme employed by CSFQ. A node that changes the DPS state carried by a packet from consistent into inconsistent is said to misbehave. In this chapter, we use the term of *node* for both a router and an end-host.

A misbehaving node can affect a subset or all flows that traverse the node. As an example, an end-host or an egress router of an ISP may intentionally modify state information carried by the packets of a *subset* of flows hoping that these flows will get a better treatment while traversing a down-stream ISP. In contrast, a router that experiences a malfunction may affect *all* flows by randomly dropping their packets.

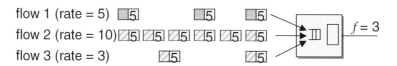

Fig. 7.1. Three flows arriving at a CSFQ router: flow 1 is consistent, flow 2 is downward-inconsistent, and flow 3 is upward-inconsistent.

A flow whose packets carry inconsistent information is called *inconsistent*; otherwise it is called *consistent*. We differentiate between two types of inconsistent flows. If the packets of a flow carry a rate smaller than the actual flow rate, we say that the flow is *downward-inconsistent*. Similarly, if the packets carry a rate that is larger than the actual flow rate we say that the flow is *upward-inconsistent*. Figure 7.1 shows an example involving three flows arriving at a CSFQ core router: flow 1 is consistent, flow 2 is downward-inconsistent, as its arrival rate is 10, but its packets carry an estimated rate of only 5, and flow 3 is upward-inconsistent since it has an arrival rate of 3, but its packets carry an estimated rate of 5. As we will show in the next section, of the two types of inconsistent flows, the downward-inconsistent ones are more dangerous as they can steal bandwidth from the consistent flows. In contrast, upward-inconsistent flows can only hurt themselves.

In summary, we assume *only* node failures that result in forwarding packets with inconsistent state. We do *not* consider general failures such as a node writing a packet IP header, e.g., spoofing the IP destination or/and source addresses, or dropping all packets of a flow.

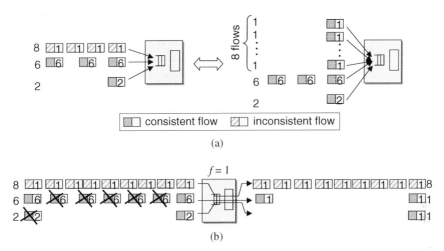

Fig. 7.2. (a) A CSFQ core router cannot differentiate between an inconsistent flow with an arrival rate of 8, whose packets carry an estimated rate of 1, and 8 consistent flows, each having an arrival rate of 1. (b) Since CSFQ assumes implicitly that all flows are consistent it will allocate a rate of 8 to the inconsistent flow, and a rate of 1 to consistent flows. The crosses indicate dropped packets.

7.1.1 Example

In this section, we first illustrate the impact of an inconsistent flow on other consistent flows that share the same link. In particular, we show that a downward-inconsistent flow may deny the service to consistent flows. Then we illustrate the impact that a misbehaving router can have on the traffic in the entire domain.

Consider a basic scenario in which three flows with rates of 8, 6, and 2 Mbps, respectively, share a 10 Mbps link. According to Eq. 4.1, the fair rate in this case is 4 Mbps[1]. As a result, the first two flows get 4 Mbps each, while flow 3 gets exactly 2 Mbps.

Next, assume that the first flow is downward-inconsistent. In particular, its packets carry an estimated rate of 1 Mbps, instead of 8 Mbps. It is easy to see then that such a scenario will break CSFQ. Intuitively, this is because a core router cannot differentiate – based only on the information carried by the packets – between the case of an 8 Mbps inconsistent flow, and the case of 8 consistent flows sending at 1 Mbps each (see Figure 7.2(a)). In fact, CSFQ will

---

[1]This is obtained by solving the equation: $\min(\alpha, 8) + \min(\alpha, 6) + \min(\alpha, 2) = 10$.

assume by default that the information carried by all packets is consistent, and, as a result, will compute a fair rate of 1 Mbps (see Figure 7.2(b)).[2] Thus, while the other two flows get 1 Mbps each, the inconsistent flow will get 8 Mbps!

Worse yet, a misbehaving router can affect not only the traffic it forwards, but also the traffic of other down-stream routers. Consider the example in Figure 7.3(a) in which the black router on the path of flow 1 misbehaves by under-estimating the rate of flow 1. As illustrated by the previous example, this will cause down-stream routers to unfairly allocate more bandwidth to flow 1, hurting in this way the consistent traffic. In this example, flow 1 will affect both flows 2 and 3. In contrast, in a stateful network, in which each router implements Fair Queueing, a misbehaving router can hurt *only* the flows it forwards. In the scenario shown in Figure 7.3(b), the misbehaving router will affect only flow 1, while the other two flows will not be affected.

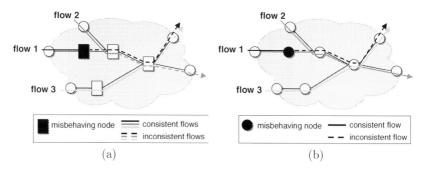

(a)   (b)

Fig. 7.3. (a) An example illustrating how a misbehaving router (represented by the black box) can affect the down-stream consistent traffic in the case of CSFQ. In particular, the misbehaving router will affect flow 1, which in turn affects flows 2 and 3 as they share the same down-stream links with flow 1. (b) In the case of Fair Queueing the misbehaving router will affect *only* flow 1; the other two flows are not affected.

7.2 The "Verify-and-Protect" Approach

To address the robustness and improve the scalability of the SCORE architecture we consider a "verify-and-protect" extension of this architecture. We achieve scalability by pushing the complexity all the way to end-hosts and eliminate the concept of the core-edge distinction. To address the trust and robustness issues, all routers statistically verify that the incoming packets

---

[2]This is obtained by solving the equation: $\min(2,\alpha)+\min(6,\alpha)+8\times\min(1,\alpha) = 10$.

are correctly marked. This approach enables routers to discover and isolate misbehaving end-hosts and routers.

The "verify-and-protect" extension consists of three components: (1) *identification* of the misbehaving nodes, (2) *protection* of the consistent traffic against the inconsistent traffic forwarded by the misbehaving node, and (3) *recovery* from the protection mode if the misbehaving node heals. Next, we briefly discuss these components in more detail.

7.2.1 Node Identification

Node identification builds on the fact that with DPS a core router can easily *verify* whether a flow is inconsistent or not. This can be simply done by having a router (1) monitor a flow, (2) re-construct its state, and then (3) check whether the reconstructed state matches the state carried by the flows' packets. We call this procedure *flow verification*. In the case of CSFQ, flow verification consists of re-estimating a flow's rate and then comparing it against the estimated rate carried by the flow's packets. If the two rates are within some predefined distance from each other (see Section 7.3.2) we say that the flow is consistent; otherwise, we say that the flow is inconsistent.

Since in our case a misbehaving node is defined as a node which forwards packets carrying inconsistent state, a simple identification algorithm would be to have each router monitor the incoming flows. Then, if the router detects an inconsistent flow, it will conclude that the up-stream node misbehaves.[3]

The drawback of this approach is that it requires core routers to monitor each incoming flow, which will compromise the scalability of our architecture. To get around this problem, we limit the number of flows that are monitored to a small sub-set of all arriving flows. While this approach does not guarantee that every inconsistent flow is identified, it is still effective in detecting misbehaving nodes. This is because, at the limit, identifying one inconsistent flow is enough to conclude that *an* up-stream node misbehaves. However, note that in this case we can no longer be certain that the up-stream neighbor misbehaves; it can be the case that another up-stream node misbehaves but the intermediate nodes fail to identify it.

7.2.2 Protection

Once a router identifies a misbehaving flow, the next step is to protect the consistent traffic against this flow. One approach would be to penalize the inconsistent flows *only*. The problem with this approach is that it is necessary

---

[3]For simplicity, here we assume that we can decide that a node misbehaves based on a *single* inconsistent flow. However, as we will show in Section 7.3.2, in the case of CSFQ we have to perform more than one flow tests to accurately identify a misbehaving node.

to maintain state for all inconsistent flows. If the number of inconsistent flows is large, this approach will compromise the scalability of the core routers.

An second approach would be to penalize *all* flows which arrive from a misbehaving node. In particular, once a router concludes that an up-stream node misbehaves, it penalizes all flows that are coming from that node, no matter whether they are inconsistent or not. While this approach may seem overly conservative, it is consistent with our failure model, which considers only node, not flow, failure.

Finally, a third approach would be to announce the failure at a higher administrative level — for example, to the network administrator — that can then take the appropriate action. At the limit, the network administrator can simply shut-down the misbehaving router and reroute the traffic. A variation of this scheme would be to design a routing protocol that automatically reroutes the traffic when a misbehaving router is identified.

In the example studied in this chapter, i.e., in the case of CSFQ, we assume the second approach (see Section 7.5).

7.2.3 Recovery

In many cases the failure of a node can be transient, i.e., after forwarding mis-behaving traffic for a certain time, a node may stop doing so. In this case, the down-stream node should detect this, and stop punishing the traffic arriving from that node. Again, this can be easily implemented by using flow verification. If a router does not detect any ill-behaved flow for a predefined period of time, it can decide then that the up-stream node no longer misbehaves, and stop punishing its flows.

7.3 Flow Verification

At the basis of the "verify-and-protect" approach lies the ability to iden-tify misbehaving nodes. In turn, this builds on the ability to perform flow verification to detect whether a flow is consistent or not. In this section we describe the flow verification algorithm, and propose a test to check for flow's consistency in the case of CSFQ.

We assume that a flow is uniquely identified by its source and destination IP addresses. This makes packet classification easy to implement at very high speeds. For most practical purposes, we can use a simple hash table with the hash keys computed over the source and destination address fields in the packet's IP header.

We consider a router architecture in which the monitoring function is im-plemented at the *input* ports. Performing flow monitoring at inputs, rather than outputs, allows us to detect inconsistent flows as soon as possible, and therefore limit the impact that these flows might have on the consistent flows

traversing the router. Without loss of generality we assume an on-line verification algorithm. The pseudocode of the algorithm is shown in Figure 7.4.
We assume that an input can monitor up to M flows simultaneously. Upon a
packet arrival, we first check whether the flow it belongs to is already monitored. If not, and if less than M flows are monitored, we add the flow to the
monitoring list. A flow is monitored for an interval of length T_{mon}. At the end
of this interval the router computes an estimate of the flow rate and compares
it against the rate carried by the flow's packets. Based on this comparison,
the router decides whether the flow is consistent or not. We call this test the
flow identification test.

upon packet p arrival:
    // if flow to which p belong is not monitored,
    // and monitoring list is not full, start to monitor it
    $f = $ get_flow_filter(p);
    if ($f \notin$ monitoring_list)
      if (size(monitoring_list) $< M$)
        // start to monitor f
        insert(monitoring_list, f);
        init($f.rate$);
        $f.start\_time = f.crt\_time$;
    else *// update f state*
    update($f.state, p$);
    if ($f.crt\_time - f.start\_time \geq T_{mon}$)
      flow_id_test($f.state, p.state$);
      ...
      delete(monitoring_list, f);

Fig. 7.4. The pseudocode of the flow verification algorithm.

A flow that fails this test is classified as inconsistent. In an ideal fluid
flow system a flow would be classified as consistent if the rate estimated
at the end of the monitoring interval is equal to the rate carried by the
flow's packets. Unfortunately, in a real system such a simple test will fail
due to inaccuracies introduced by (1) the rate estimation algorithm, (2) the
delay jitter, and (3) the probabilistic buffer management scheme employed
by CSFQ. In Section 7.3.2 we present a flow identification test that is robust
in the presence of these inaccuracies.

One important question is whether the need to maintain state for each
flow that is verified does not compromise the scalability of our approach. We
answer this question next. Let T_{mon} be the average time it takes a router to
verify a flow (see Table 7.1). Since according to the algorithm in Figure 7.4,
a new flow (to be verified) is selected by choosing a random packet, and a
router can verify up to M flows simultaneously, the expected time to select

Notation	Comments
M	maximum number of flows simultaneously monitored at an input port
T_{mon}	monitoring interval
T_{inc}	expected time to identify a flow as inconsistent
n_{inc}	expected number of tests to classify a flow as inconsistent
k_{ov}	*overflow factor* – ratio between labels carried by flow's packets at the entrance of the network and the fair rate on the upstream bottleneck link
k_{inc}	*inconsistency factor* – ratio between flow's arrival rate and labels carried by flow's packets
m	number of packets sent during T_{mon} at the fair rate (assuming fixed packet sizes)
S_c	event that a tested flow is *consistent*
S_{inc}	event that a tested flow is *inconsistent*; $\Pr(S_{inc}) + \Pr(S_c) = 1$
C_c	event that a tested flow is classified as *consistent*
C_{inc}	event that a tested flow is classified as *inconsistent*
$p_{c-inc} = \Pr(C_{inc} \mid S_c)$	probability that a consistent flow is misidentified
$p_{inc-inc} = \Pr(C_{inc} \mid S_{inc})$	probability that an inconsistent flow is identified
p_a	probability that a selected flow is active long enough to be tested
$p_{inc} = p_a \Pr(C_{inc})$	probability that a selected flow is classified as inconsistent
$p_{id} = \Pr(S_{inc} \mid C_{inc})$	probability to identify an inconsistent flow, i.e., probability that a flow classified as inconsistent is indeed inconsistent
f_{inc}	fraction of inconsistent traffic
$\Pr(S_{inc})$	probability to select an inconsistent flow; we assume $\Pr(S_{inc}) = f_{inc}$

Table 7.1. Notations used throughout this chapter. For simplicity, the notations do not include the time argument t.

an inconsistent flow is about $T_{mon} / (M f_{inc})$, where f_{inc} represents the fraction of the inconsistent traffic (see Table 7.1). Thus, the expected time to eventually catch an inconsistent flow is $T_{mon}/(M f_{inc}) + T_{mon}$. As a result, it makes little sense to choose M much larger than $1/f_{inc}$, as this will only marginally reduce the time to catch a flow. In fact, if we choose $M \simeq 1/f_{inc}$, the expected time to eventually catch an inconsistent flow is within a factor of two of the optimal value T_{mon}. Next, it is important to note that in practice we can ignore the inconsistent traffic, when f_{inc} is small. Indeed, in the worst case, ignoring the inconsistent traffic is equivalent to "losing" only a fraction f_{inc} of the link capacity to the inconsistent traffic. For example, if

$f_{inc} = 1\%$, even if we ignore the inconsistent traffic, the consistent traffic will still receive about 99% of the link capacity. Given the various approximations in our algorithms, ignoring the inconsistent traffic when f_{inc} is on the order of a few percent is acceptable in practice. As a result, we expect that the number of flows that are simultaneously monitored, M, to be on the *order of tens*. Note that M does not depend on the number of flows that traverse a router, a number that can be much larger, i.e., on the order of hundred of thousands or even millions.

7.3.1 Bufferless Packet System

To facilitate the discussion of the flow identification test, we consider a simplified buffer-less packet system, and ignore the inaccuracies due to the rate estimation algorithm. For simplicity, assume that each source sends constant-bit rate traffic, and that all packets have the same length. We consider only end-host and/or router misbehaviors that result in having *all* packets of a flow carry a label that is k_{inc} times *smaller* than the actual flow rate, where $k_{inc} \neq 1$. Thus, we assume that k_{inc}, also called *inconsistency factor*, does not change[4] during the life of a flow. In Figure 7.1, flow 2 has $k_{inc} = 2$, while flow 3 has $k_{inc} = 3/5$.

We consider a congested link between two routers N_1 and N_2, and denote it by $(N_1{:}N_2)$. We assume that N_1 forwards traffic to N_2, and that N_2 monitors it. Let α be the fair rate of $(N_1{:}N_2)$. Then, with each flow that arrives at N_1 and which is forwarded to N_2 along $(N_1{:}N_2)$, we associate an *overflow factor*, denoted k_{ov}, that represents the ratio between the labels carried by the flow's packets and the fair rate α along $(N_1{:}N_2)$. For example, in Figure 7.1 each flow has $k_{ov} = 5/3$.

7.3.2 Flow Identification Test

In this section we present the flow identification test for CSFQ in the buffer-less model. First, we show why designing such a test is difficult. In particular, we demonstrate that the probabilistic dropping scheme employed by CSFQ can significantly affect the accuracy of the test. Then we give three desirable goals for the identification test, and discuss the trade-offs to achieve these goals.

Our flow identification test makes the decision based on the *relative discrepancy* between the rate of the flow estimated by the router, denoted $\hat{r}$, and the labels carried by the flow's packets, denoted $\bar{r}$. More precisely, the relative discrepancy is defined as

---

[4]The reason for this assumption is that a constant value of k_{inc} maximizes the excess service received by an inconsistent flow without increasing the probability of the flow to be caught. We show this in Section 7.4.1.

$$dis_{rel} = \frac{\widehat{r} - \bar{r}}{\bar{r}}. \tag{7.1}$$

In an idealized fluid flow system the flow identification test needs only to see whether the relative discrepancy is zero or not. Unfortunately, even in our simplified buffer-less model this test is not good enough, primarily due to the inaccuracies introduced by the CSFQ probabilistic dropping scheme. Even if a flow is consistent and it traverses only well-behaved nodes, it can still end-up with a non-zero discrepancy. To illustrate this point consider a consistent flow with the overflow factor $k_{ov} > 1$ that arrives at N_1. Let r be the arrival rate of the flow and let α be the fair rate of $(N_1:N_2)$. Note that $k_{ov} = r/\alpha$. Assume that exactly n packets of the flow arrive at N_1 during a monitoring interval T_{mon}, and the packet dropping probability is independently distributed. Then, the probability that N_1 will forward exactly x packets during T_{mon} is:

$$p_{fwd}(n;x) = \binom{n}{x} p^x (1-p)^{n-x} = \binom{n}{x} \left(\frac{1}{k_{ov}}\right)^x \left(1 - \frac{1}{k_{ov}}\right)^{n-x}, \tag{7.2}$$

where $p = \alpha/r = 1/k_{ov}$ represents the probability to forward a packet.

Next, assume that N_2 monitors this flow. Since we ignore the delay jitter, the probability that N_2 receives exactly x packets during T_{mon} is exactly the probability that N_1 will forward x packets during T_{mon}, i.e., $p_{fwd}(n;x)$. As a result, N_2 estimates the flow rate as $\widehat{r} = xl/T_{mon}$ with probability $p_{fwd}(n;x)$, where l is the packet length. Let m denote the number of packets that are forwarded at the fair rate α, i.e., $m = \alpha T_{mon}/l$. The relative discrepancy of the flow measured by N_2 during T_{mon} is then:

$$dis_{rel} = \frac{\widehat{r} - \bar{r}}{\bar{r}} = \frac{\widehat{r} - \alpha}{\alpha} = \frac{x - m}{m}. \tag{7.3}$$

with the probability $p_{fwd}(n;x)$.

Figure 7.5 depicts the probability density function (p.d.f.) of dis_{rel} for $m = 10$ and different values of k_{ov}. Thus, even if the flow is consistent, its relative discrepancy as measured by N_2 can be significant. This suggests a range based test to reduce the probability of false positives, i.e, the probability a consistent flow will be classified as being inconsistent. In particular, we propose the following flow identification test:

Flow Identification Test (CSFQ) *Define two thresholds: $H_l < 0$, and $H_u > 0$, respectively. Then we say that a flow is inconsistent if its relative discrepancy (dis_{rel}) is either smaller than H_l, or larger than H_u.*

As shown in Figure 7.5, the overflow factor (k_{ov}) has a significant impact on the flow's relative discrepancy (dis_{rel}). The higher the overflow factor of a flow is (i.e., the more aggressive a flow is), the more spread out its relative discrepancy is. A spread out relative discrepancy makes it more difficult to

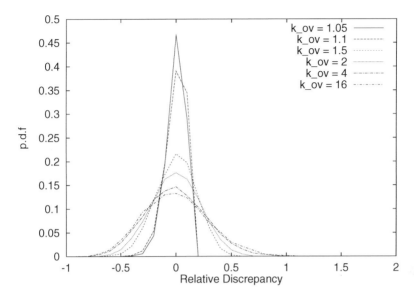

Fig. 7.5. The probability density function (p.d.f.) of the relative discrepancy of estimating the flow rate for different values of k_{ov}

accurately identify an inconsistent flow. Assume $H_u = 0.5$, that is, a flow will be classified as inconsistent whenever the measured relative discrepancy (dis_{rel}) exceeds 0.5. As shown in Figure 7.5, in this case, the probability a consistent flow will be misidentified, increases significantly with k_{ov}. If $k_{ov} \leq 1.05$, this probability is virtually 0, while if $k_{ov} = 16$, this probability is about 0.03, which in practice can be unacceptable. Note that while we can decrease this probability by increasing H_u, such a simple solution has a major drawback: a large H_u will allow flows with a larger inconsistency factor, i.e., with $k_{inc} \leq H_u + 1$, to go by undetected.

To simplify the problem of choosing H_u, we assume that the overflow factor of a consistent flow has an upper bound. This assumption is motivated by the observation that consistent flows are likely to react to congestion, and therefore their overflow factor will be small. Indeed, unless a flow is malicious, it makes little sense for the source to send at a rate higher than the available rate on the bottleneck link. As a result, we expect that k_{ov} to be slightly larger than 1 in the case of a consistent flow. For this reason, in the remaining of this chapter we assume that $k_{max} \leq 1.3$. The value of 1.3 is chosen somewhat arbitrary to coincide to the value used in [101] to differentiate between well-behaved and malicious (overly aggressive) flows. Thus, if a consistent flow is too aggressive, i.e., its overflow factor $k_{ov} \geq 1.3$, it will run a high risk of being classified as inconsistent. However, we believe that this is the right tradeoff, since it will provide an additional incentive to end-hosts to implement flow congestion control.

7.3.3 Setting Threshold H_u

The main question that remains to be answered is how to set up thresholds H_u and H_l. In the remainder of this section we give some guidelines and illustrate the trade-offs in choosing H_u. Since choosing H_l faces similar trade-offs, we do not discuss this here.

Setting H_u is difficult because the flow identification test has to meet several conflicting goals:

1. *robustness* - maximize the probability that a flow identified as inconsistent is indeed inconsistent. We denote this probability by p_{id}.
2. *sensitivity* - minimize the inconsistency factor (k_{inc}) for which a flow is still caught.
3. *responsiveness* - minimize the expected time it takes to classify a flow as inconsistent. As a metric, we consider the expected number of tests it takes to classify a flow as inconsistent. We denote this number by n_{inc}.

In the remainder of this section, we assume that router N_1 misbehaves by forwarding a constant fraction of inconsistent traffic f_{inc} to N_2. The reason we assume that f_{inc} is constant is because, as we will show in Section 7.4.1, this represents the worst case scenario for our flow identification test. In particular, if a malicious router wants to maximize the excess bandwidth received by its inconsistent traffic before being caught, then it has to send inconsistent traffic at a constant fraction of its total traffic. Further, we assume that all consistent flows have $k_{ov} = 1.3$, and all inconsistent flows have $k_{ov} \leq 1$, i.e., no packet of an inconsistent flow is ever dropped by N_1. As a result the relative discrepancy of an inconsistent flow will be exactly k_{inc}. This means that our test will not be able to catch an inconsistent flow with $k_{inc} \leq H_u + 1$. Again, this choice of k_{ov} represents the worse-case scenario. It can be shown that decreasing the k_{ov} of consistent flows and/or increasing the k_{ov} of inconsistent flows will only improve the test robustness, sensitivity, and responsiveness.

Next, we derive the two parameters that characterize the robustness and responsiveness: p_{id}, and n_{inc}. By using Bayes's formula and the notations in Table 7.1 we have

$$
\begin{aligned}
p_{id} &= \Pr(S_{inc} \mid C_{inc}) \\
&= \frac{\Pr(S_{inc})\Pr(C_{inc}|S_{inc})}{\Pr(S_{inc})\Pr(C_{inc}|S_{inc}) + \Pr(S_c)\Pr(C_{inc}|S_c)} \\
&= \frac{f_{inc} \times p_{inc-inc}}{f_{inc} \times p_{inc-inc} + (1 - f_{inc}) \times p_{c-inc}}.
\end{aligned}
\tag{7.4}
$$

In addition, the expected number of flows that are tested before a flow will be classified as inconsistent, n_{inc}, is

$$n_{inc} = \sum_{i \geq 1} i(1 - p_{inc})^{i-1} p_{inc} = \frac{1}{p_{inc}}, \qquad (7.5)$$

where p_{inc} is the probability that a selected flow will be classified as inconsistent. With the notations from Table 7.1, and using simple probability manipulations, we have

$$\begin{aligned}
p_{inc} &= p_a \Pr(C_{inc}) \qquad\qquad\qquad\qquad\qquad\qquad\qquad (7.6) \\
&= p_a(\Pr(C_{inc} \cap S_{inc}) + \Pr(C_{inc} \cap S_c)) \\
&= p_a(\Pr(S_{inc})\Pr(C_{inc} \mid S_{inc}) + \Pr(S_c)\Pr(C_{inc} \mid S_c)) \\
&= p_a \times (f_{inc} \times p_{inc-inc} + (1 - f_{inc}) \times p_{c-inc})
\end{aligned}$$

Finally, by combining Eqs. (7.5) and (7.6), we obtain

$$n_{inc} = \frac{1}{p_a \times (f_{inc} \times p_{inc-inc} + (1 - f_{inc}) \times p_{c-inc})}. \qquad (7.7)$$

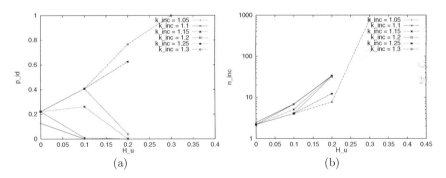

Fig. 7.6. (a) The probability to identify an inconsistent flow, p_{id}, and (b) the expected number of tests it takes to classify a flow as inconsistent, n_{inc}, as functions of H_u. (The values of n_{inc} for $k_{inc} \leq 1.25$ and $H_u = 0.3$ are not plotted as they are larger than 10^6.) All inconsistent flows have $k_{ov} = 1$, $f_{inc} = 0.1$, and $m = 10$.

As illustrated by Eqs. (7.4) and (7.7), the fraction of the inconsistent traffic, f_{inc}, has a critical impact on both p_{id} and n_{inc}. The smaller f_{inc} is, the smaller p_{id} is, and the larger n_{inc} is. The reason n_{inc} increases when f_{inc} decreases is because in any valid flow identification test we have $p_{inc-inc} > p_{c-inc}$, i.e., the probability an inconsistent flow will be identified is always larger than the probability a consistent flow will be misidentified.

In the remainder of this chapter, we choose a somewhat arbitrary $f_{inc} = 0.1$. If $f_{inc} < 0.1$, we simply ignore the impact of the inconsistent traffic on the consistent traffic. This decision is motivated by the fact that the inconsistent traffic can "steal" at most a fraction, f_{inc}, of the link capacity. In the worst case, this leads to a 10% degradation in the bandwidth received by a consistent flow, which we consider to be acceptable. However, it should be noted that there is nothing special about this value of f_{inc}. The reason for which we use a specific value for f_{inc} is to make the procedure of choosing H_u more concrete.

Without loss of generality, we assume $p_a = 1$, i.e., once a flow is selected, it remains active for at least T_{mon}. Note that if $p_a < 1$, this will simply result in scaling up n_{inc} by $1/p_a$. Probabilities $p_{inc-inc}$ and p_{c-inc} are computed based on the p.d.f. of the relative discrepancy. Figure 7.6(a) plots then the probability to identify an inconsistent flow, p_{id}, while Figure 7.6(b) plots the expected number of tests to classify a flow as inconsistent, n_{inc}. These plots illustrate the trade-offs in choosing H_u. On one hand, the results in Figure 7.6(a) suggest that we have to choose $H_u \geq 0.2$; otherwise, the probability to identify an inconsistent flow becomes smaller than 0.5. On the other hand, from Figure 7.6(b), it follows that in order to make the test responsive we have to choose $H_u \leq 0.2$, as n_{inc} starts to increase hyper-exponentially for $H_u > 0.2$. To meet these restrictions, we choose $H_u = 0.2$. Note that this choice gives us the ability to catch any inconsistent flow as long as its inconsistency factor (k_{inc}) is greater than 1.2.

7.3.4 Increasing Flow Identification Test's Robustness and Responsiveness

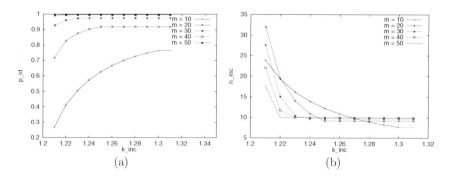

Fig. 7.7. (a) The probability to identify an inconsistent flow, p_{id}, and (b) The expected number of tests to classify a flow as inconsistent, n_{inc}, versus inconsistency factor, k_{inc}, for various values of m.

In the previous section we have assumed that $m = 10$, where m represents the number of packets that can be sent during a monitoring interval at the fair rate by the upstream router N_1. In this section we study the impact of m on the flow identification test performances.

Figures 7.7(a) and 7.7(b) plot the probability to identify an inconsistent flow, p_{id}, and the expected number of tests it takes to classify a flow as inconsistent, n_{inc}, versus k_{inc} for various values of m. As shown in Figure 7.7(a), increasing m can dramatically improve the test robustness. In addition, as shown in Figure 7.7(b), a large m makes the identification test more responsive for values of k_{inc} closer to H_u. However, it is important to note that, while a large m can significantly reduce n_{inc}, this does not necessary translate into a reduction of the *time* it takes to classify a flow as inconsistent, i.e., T_{inc}. In particular, if a router monitors M flows simultaneously, we have $T_{inc} = T_{mon} n_{inc}/M$. Thus, an increase of m results directly into an increase of the time it takes to test a flow for consistency, T_{mon}, an increase which can offset the decrease of n_{inc}.

7.4 Identifying Misbehaving Nodes

Recall that our main goal is to detect misbehaving routers or/and hosts. In this section we present and discuss such a test.

Like a flow identification test, ideally, a *node* identification test should be (1) robust, (2) responsive, and (3) sensitive. Unfortunately, it is very hard, if not impossible, to achieve these goals simultaneously. This is to be expected as the node test is based on the flow identification test, and therefore we are confronted with the same difficult trade-offs. Worse yet, the fact that the probability of a node misbehaving can be very low makes the problem even more difficult.

Arguably, the simplest node identification test would be to decide that an upstream node misbehaves whenever a flow is classified as inconsistent. The key problem is that in a well engineered system, in which the probability of a misbehaving node is very small, this test is not enough. As an example, assume that we use $m = 20$, and that our goal is to detect inconsistent flows with $k_{inc} \geq 1.25$. Then, according to the results in Figure 7.7, we have $p_{id} = 0.92$. Thus, there is a 0.08 probability of false positives. In addition, assume that the probability that the upstream node N_1 misbehaves is also 0.08. This basically means that whenever N_2 classifies a flow as being inconsistent there is only a 0.5 chance that this is because N_1 misbehaves!

To alleviate this problem we propose a slightly more complex node identification test: instead of using only one observation to decide whether a node misbehaves or not, we use multiple observations. In particular we have the following test

Node Identification (CSFQ): Test 1 *An upstream node is assumed to misbehave if at least n_t flows out of the last N_t tested flows were classified as inconsistent, where n_t and N_t are predefined constants.*

Let P denote the probability of false positives, that is, the probability that every time we identify an inconsistent flow during the last N_t tests we were wrong. Assuming that the results of the flow identification tests are independently distributed, we have:

$$P = 1 - \sum_{i=n_t}^{N_t} \binom{N_t}{i} (1 - p_{id})^i p_{id}^{N_t - i}. \tag{7.8}$$

In the context of the previous example, assume that $N_t = 5$, and $n_t = 3$. This yields $P = 0.0002$. Thus, in this case, the probability that we are wrong is much smaller than the probability that the node is misbehaving. In consequence, if three out of the last five tested flows are classified as inconsistent we can decide with high probability that the upstream node is indeed misbehaving.

A potential problem with the previous test is that a large threshold H_u, will allow a misbehaving node to inflict substantial damages without being caught. In particular, if a node inserts in the packet headers a rate that is k_{inc} times larger than the actual flow rate, where $k_{inc} - 1 < H_u$, the node will get up to $k_{inc} - 1$ times more capacity for free without being caught. For example, in our case, when $H_u = 1.2$, the upstream node can get up to 20% extra bandwidth.

To filter out this attack we employ a second test. This test is based on the observation that in a system in which all nodes are well-behaved the mean value of dis_{rel} is expected to be zero. In this case, any significant deviation of dis_{rel} from zero is interpreted as being caused by a misbehaving upstream node.

Node Identification (CSFQ): Test 2 *An upstream node is assumed to misbehave if the mean value of dis_{rel} does not fall within $[-\delta, \delta]$.*

7.4.1 General Properties

In this section we present two simple but important properties of our identification test. The first property says that a misbehaving node can inflict maximum of damage when it sends inconsistent traffic at a constant rate. In particular, the excess bandwidth received by the inconsistent traffic at a downstream router, before being caught, is maximized when the rate of the inconsistent traffic is constant. This property is important because it allows us to limit our study to the cases in which a the fraction of the inconsistent traffic sent by a misbehaving node is constant. The second property says that a misbehaving node cannot hurt the downstream consistent traffic in a "big" way for a long time. In other words, the higher the rate of the inconsistent traffic is, the faster the misbehaving node is caught. The two properties are given below.

Property 7.1. Let $R_{inc}(t', t'')$ denote the total volume of inconsistent traffic received by a router at an input port during the interval $[t', t'')$. Then, the probability that no inconsistent flow is identified during $[t', t'')$, is minimized when the inconsistent traffic arrives at a fixed rate $\overline{r_{inc}}$, where $\overline{r_{inc}} = R_{inc}(t', t'')/(t'' - t')$.

Proof. Let $r_{inc}(t)$ be the rate of the inconsistent traffic at time t, and let C denote the input link capacity where the traffic arrives. Then, the fraction of the inconsistent traffic at time t is $f_{inc}(t) \geq r_{inc}(t)/C$, where the equality occurs when the link is fully utilized. Let $t_1, t_2, \ldots, t_n$ be the time instants when a new flow is selected to be monitored during the interval $[t', t'')$. By using Eq. (7.6), the probability that none of these flows will be identified as being inconsistent, denoted q, is

$$
\begin{aligned}
q &= \prod_{i=1}^{n}(1 - p_{inc}(t_i)) \qquad\qquad\qquad\qquad\qquad (7.9)\\
&\leq \left(1 - p_a \frac{\sum_{i=1}^{n}(f_{inc}(t_i) \times p_{inc-inc} + (1 - f_{inc}(t_i)) \times p_{c-inc})}{n}\right)^n \\
&= \left(1 - p_a \frac{\sum_{i=1}^{n} f_{inc}(t_i)}{n} p_{inc-inc} - p_a \frac{\sum_{i=1}^{n} 1 - f_{inc}(t_i)}{n} p_{c-inc}\right)^n.
\end{aligned}
$$

Since a router randomly selects a flow to be monitored, we assume without loss of generality that $t_1, t_2, \ldots, t_n$ are independently distributed within $[t', t'')$. By assuming that $r_{inc}(t)$ is a continuous function over the interval $[t', t'')$, we have

$$
\overline{r_{inc}} = \text{Exp}\left(\frac{\sum_{i=1}^{n} r_{inc}(t_i)}{n}\right). \qquad\qquad\qquad (7.10)
$$

In addition, since $f_{inc}(t) \geq r_{inc}(t)/C$, for any t, the following inequality follows trivially

$$
\frac{\sum_{i=1}^{n} f_{inc}(t_i)}{n} \geq \frac{\sum_{i=1}^{n} r_{inc}(t_i)}{nC}. \qquad\qquad\qquad (7.11)
$$

By combining Eqs. (7.9), (7.11) and Eq. (7.10), using the fact that f_{inc}, p_a, $p_{inc-inc}$ and p_{c-inc} are independently distributed variables, and the fact that $p_{inc-inc} > p_{c-inc}$, we obtain

$$\text{Exp}(q) \leq \left(1 - \text{Exp}(p_a)\frac{\overline{r_{inc}}}{C}\text{Exp}(p_{inc-inc}) - \text{Exp}(p_a)\left(1 - \frac{\overline{r_{inc}}}{C}\right)\text{Exp}(p_{c-inc})\right)^n$$

$$= (1 - \text{Exp}(p_a)\,\overline{f_{inc}}\,\text{Exp}(p_{inc-inc}) - \text{Exp}(p_a)\,(1 - \overline{f_{inc}})\,\text{Exp}\,(p_{c-inc}))^n$$

where $\overline{f_{inc}} = \overline{r_{inc}}/C$. But the last term in the above inequality represents exactly the expected probability a flow will not be classified as inconsistent after n tests, when the inconsistent traffic arrives at a fixed rate. This concludes the proof of the property. $\square$

Thus, if the fraction at which a misbehaving node sends inconsistent traffic fluctuates, the probability to identify an inconsistent flow can only increase. This is exactly the reason we have considered in this chapter only the case in which this fraction, i.e., f_{inc}, is constant.

Property 7.2. The higher the rate of the inconsistent traffic is, the higher the probability to identify an inconsistent flow is.

Proof. The proof follows immediately from Eq. (7.4) and the fact that in a well designed flow identification test, the probability an inconsistent flow will be identified, $p_{inc-inc}$, should be larger than the probability a consistent flow will be misidentified, p_{c-inc}. $\square$

7.5 Simulation Results

In this section we evaluate the accuracy of the flow identification test by simulation. All simulations were performed in ns-2 [78], and are based on the original CSFQ code available at http://www.cs.cmu.edu/~istoica/csfq.

Unless otherwise specified, we use the same parameters as in Section 4.4. In particular, each output link has a buffer of 64 KB, and the buffer threshold for CSFQ is set to 16 KB. The averaging constant to estimate the flow rate is $K = 100$ ms, and the averaging constant to estimate the fair rate is $K_\alpha = 200$ ms. In all topologies we assume that the first router traversed by each flow estimates the rate of the flow and inserts it in the packet headers.

Each router can verify up to four flows simultaneously. Each flow is verified for at least 200 ms or 10 consecutive packets. Core routers use the same algorithm as edge routers to estimate the flow rate. However, to improve the convergence of the estimation algorithm, the rate is initialized to the label of the first packet of the flow that arrives during the estimation interval. As discussed in Section 7.3.3, we set the upper threshold to $H_u = 1.2$. Since we will test only for downward-inconsistent flows – the only type of inconsistent flows that can steal service from consistent flows – we will not use the lower threshold H_l.

In general, we find that our flow identification test is robust and that our results are quite comparable to the results obtained in the case of the bufferless system model.

7.5.1 Calibration

As discussed in Section 7.3.2, to design the node identification test – i.e., to set parameters n_t and N_t – it is enough to know (1) the probability to identify an inconsistent flow, p_{id}, and (2) the expected number of tests it takes to classify a flow as inconsistent, n_{inc}.

So far, in designing our identification tests, we have assumed a bufferless system, which ignores the inaccuracies introduced by (1) the delay jitter, and (2) the rate estimation algorithm. In this section we simulate a more realistic scenario by using the ns-2 simulator [78]. We consider a simple link traversed by 30 flows out of which three are inconsistent and 27 are consistent. Note that this corresponds to a $f_{inc} \simeq 0.1$. Our goal is twofold. First we want to show that the results achieved by simulations are reasonably close to the ones obtained by using the bufferless system. This basically says that the bufferless system can be used as a reasonable first approximation to design the identification tests. Second, we use the results obtained in this section to set parameters N_t and n_t for the node identification test.

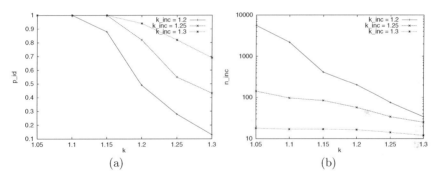

(a)   (b)

Fig. 7.8. (a) The probability to identify an inconsistent flow, p_{id}, and (b) the expected number of tests it takes to classify a flow as inconsistent, n_{inc}. We consider 30 flows, out of which three are inconsistent.

We perform two experiments. In the first experiment, we assume that all flows are constant bit rate UDPs. Figure 7.8 (a) plots the probability p_{id}, while Figure 7.8 (b) plots the expected number of flows that are tested before a flow is classified as inconsistent. As expected, when $k_{inc} < 1.2$, $p_{id} \rightarrow 1$. This is because even in the worst-case scenario, when all packets of a consistent flow are forwarded, the flow's relative discrepancy will be no larger than $H_u = 0.2$. However, as k_{inc} exceeds 1.2, the probability p_{id} reduces significantly, as we are more and more likely to classify a consistent flow as inconsistent. In addition, p_{id} is strongly influenced by k_{inc}. The larger the k_{inc} is, the larger p_{id} is. This is because when k_{inc} increases we are more and more

k_{inc}	p_{id}	n_{inc}
1.3	0.76	25
1.25	0.45	63.6
1.2	0.12	115
1.15	0.025	690
1.1	0	700
1	0	720

Table 7.2. p_{id} and n_{inc} as a function of the parameter k_{inc} of the inconsistent flows. We consider 27 consistent TCP flows and 3 UDP inconsistent flows.

likely to catch the inconsistent flows. Similarly, as shown in Figure 7.8 (b), n_{inc} is decreasing in k_{inc} and increasing in k_{inc}.

One thing worth noting here is that these results are consistent with the ones achieved in the simplified bufferless system. For example, as shown in Figures 7.6(a) and 7.8(a), in both cases the probability p_{id} for $k_{inc} = 1.3$, $k_{inc} = 1.25$, and $H_u = 0.7$, is somewhere between 0.7 and 0.8. This suggests that tuning the parameters in the bufferless system represents a reasonable first order approximation.

In the second experiment we assume that all the consistent flows are TCPs, while the inconsistent flows are again constant bit rate UDPs. The results are presented in Table 7.2. It is interesting to note that although the TCP flows experience a dropping rate of only 5%, which would correspond to $k_{inc} = 1.05$, both p_{id} and n_{inc} are significantly worse than their corresponding values in the case of the previous experiment when $k_{inc} = 1.05$. In fact, comparing the values which corresponds to $k_{inc} = 1.3$ in both Figure 7.8(a) and Table 7.2, the behavior of TCPs is much closer to a scenario in which they are replaced by UDPs with $k_{inc} = 1.3$, rather than $k_{inc} = 1.05$. This is mainly because the TCP burstiness negatively affects the rate estimation accuracy, and ultimately the accuracy of our identification test.

In summary, for both TCP and UDP flows we take $p_e = 0.76$ and $n_{inc} = 25$. This means that by using 25 observations, the probability to identify an inconsistent flow with $k_{inc} \geq 1.3$ is greater or equal to 0.76.

Finally, for the node identification test we choose $n_t = 2$ and $N_t = 25$. It can be shown that this choice increases the probability to correctly identify an inconsistent node to over 0.9.

7.5.2 Protection and Recovery

In this section we illustrate the behavior of our scheme during the protection and recovery phases. For this we consider a slightly more complex topology consisting of five routers (see Figure 7.9(a)). There are ten flows from router 1 to router 4, and another 10 flows from router 2 to router 4. We assume that router 1 misbehaves by forwarding an inconsistent flow with $k_{inc} > 1$. Again,

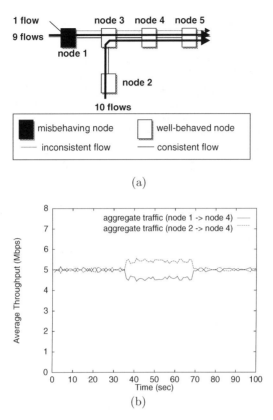

Fig. 7.9. (a) Topology used to illustrate the protection and recovery aspects of our scheme. (b) The aggregate throughput of all flows from router 1 to router 5, and of all flows from router 2 to router 5, respectively. All consistent flows are UDPs with $k_{inc} = 1.3$. The inconsistent flow is UDP and has $k_{inc} = 1.3 = k_{inc} = 1.3$.

note that in this case the fraction of the inconsistent traffic is $f_{inc} \simeq 0.1$. Finally, we assume that router 1's misbehavior is transitory, i.e., it starts to misbehave at time $t = 33$ sec, and ceases to misbehave at time $t = 66$ sec. The entire simulation time is 100 sec.

In the first experiment we assume that all consistent flows have $k_{inc} = 1.3$, and that the inconsistent flow has $k_{inc} = 1.3$, and $k_{ov} = 1.3$. Figure 7.9(b) plots the aggregate throughputs of all flows from router 1 to router 5, and from router 2 to router 5, respectively. Note that during the interval $(33sec, 66sec)$, i.e., during the time when router 1 misbehaves, the flows from router 2 to router 5 get slightly higher bandwidth. This illustrates the fact that even for $k_{inc} = 1.3$ (which is the the minimum k_{inc} we have considered in designing the node identification test), our algorithm is successful in identifying the node misbehavior, and in protecting the well-behaved traffic that flows from router 2 to router 5. In addition, once router 1 ceases to misbehave at $t = 66$ sec our recovery algorithm recognizes this and stops punishing the traffic forwarded by router 1.

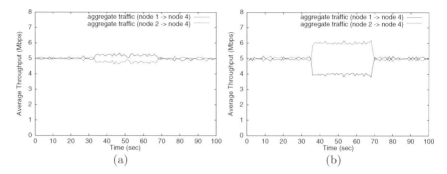

Fig. 7.10. Aggregate throughputs of all flows from router 1 to router 5, and from router 2 to router 5, when all routers implement (a) the unmodified CSFQ, and (b) the "verify-and-protect" version of CSFQ. All consistent flows are UDPs with $k_{inc} = 1.3$. The inconsistent flow is UDP with $k_{inc} = 2$, and $k_{inc} = 1$.

In Figures 7.10 (a) and (b) we plot the results for a virtually identical scenario both in the case when all routers employ the unmodified CSFQ, and in the case when the routers employ the "verify-and-protect" version of CSFQ. The only difference is that the inconsistent flow now has $k_{inc} = 2$, instead of 1.3. The reason for the increase of k_{inc} is to make it easier to observe the amount of bandwidth that is stolen by the inconsistent flow from the consistent flows when the unmodified CSFQ is used. This can be seen in Figure 7.10 (a) as the aggregate throughput of all flows from router 2 to router 5 is slightly less than 5 Mbps. In contrast, when the "verify-and-protect" version is used, these flows get considerably more than 5Mbps. This is because once router 3 concludes that router 1 misbehaves, it punishes all its flows by simply multiplying their labels by $k_{inc} \simeq 2$. As a result, the traffic traversing router 1 gets only 33%, instead of 50%, of the link capacities at the subsequent routers.

Finally, Figures 7.11 (a) and 7.11(b) plot the results for the case in which consistent flows are replaced by TCPs. As expected, the results are similar: the "verify-and-protect" version of CSFQ is able to successfully protect and recover after router 1 stops misbehaving.

7.6 Summary

While the solutions based on SCORE/DPS have many advantages over tradi-tional stateful solutions (see Chapter 4, 5, 6), they still suffer from robustness

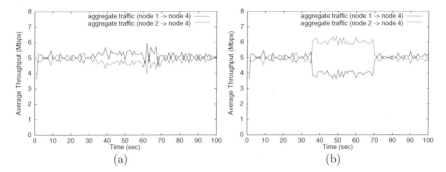

Fig. 7.11. Aggregate throughputs of all flows from router 1 to router 5, and from router 2 to router 5, when all routers implement (a) the unmodified CSFQ, and (b) the "verify-and-protect" version of CSFQ. All consistent flows are TCPs. The inconsistent flow is UDP with $k_{inc} = 2$, and $k_{inc} = 1$.

and scalability limitations when compared to the stateless solutions. In particular, the scalability is hampered because the network core cannot transcend trust boundaries (such as the ISP-ISP boundaries), and therefore high-speed routers on these boundaries must be stateful edge routers. The lack of robustness is because the malfunctioning of a single edge or core router could severely impact the performance of the entire network, by inserting inconsistent state in the packet headers.

In this chapter, we have proposed an approach to overcome these limitations. To achieve scalability we push the complexity all the way to the end-hosts. To address the trust and robustness issues, all routers statistically verify whether the incoming packets carry consistent state. We call this approach "verify-and-protect". This approach enables routers to discover and isolate misbehaving end-hosts and routers. The key technique needed to implement this approach is flow verification that allows the identification of packets carrying inconsistent state. To illustrate this approach, we have described the identification algorithms in the case of Core-Stateless Fair Queueing (CSFQ), and we have presented simulation results in ns-2 [78] to demonstrate its effectiveness.

A final observation is that the "verify-and-protect" approach can also be useful in the context of stateful solutions. Based only on the traffic a router sees, these solutions are not usually able to tell whether there is a misbehaving up-stream router or not. For example, with Fair Queueing there is no way for a router to discover an up-stream router that spuriously drops packets, as it cannot differentiate between a flow whose packets were dropped by a misbehaving router, and a flow who just sends fewer packets. Thus, the "verify-and-protect" approach can be also used to effectively increase the robustness of the traditional stateful solutions.

8 Prototype Implementation Description

In Chapters 4, 5 and 6 we have presented three applications of our SCORE/DPS solution to provide scalable services in a network in which core routers maintain no per flow state. In this chapter, we describe a prototype implementation of our solution. In particular, we have fully implemented the functionalities required to provide guaranteed and flow protection services. While section 5.5 presents several experimental results, including overhead measurements of our prototype, this chapter focuses on the implementation details.

The main goal of the prototype implementation is to show that it is possible to efficiently deploy our algorithms in today's IPv4 networks with minimum incompatibility. To prove this point we have implemented a complete system, including support for fine grained monitoring, and easy configuration. The current prototype is implemented in FreeBSD v2.2.6, and it is deployed in a test-bed consisting of 300 MHz and 400 MHz Pentium II PCs connected by point-to-point 100 Mbps Ethernets. The test-bed allows the configuration of a path with up to two core routers.

Although we had complete control of our test-bed, and, due to resource constraints, the scale of our experiments was rather small (e.g., the largest experiment involved just 100 flows), we have devoted special attention to making our implementation as general as possible. For example, while in the current implementation we re-use protocol space in the IP header to store the DPS state, we make sure that the modified fields can be fully restored by the egress router. In this way, the changes operated by the ingress and core routers on the packet header are completely transparent to the outside world. Similarly, while the limited scale of our experiments would have allowed us to use simple data structures to implement our algorithms, we try to make sure that our implementation is scalable. For example, instead of using a simple linked list to implement the CJVC scheduler, we use a calendar queue together with a two-level priority queue to efficiently handle a very large number of flows (see Section 8.1).

For debugging and management purposes, we have implemented full support for packet level monitoring. This allows us to visualize simultaneously and in real-time the throughputs and the delays experienced by flows at different points in the network. A key challenge when implementing such a fine

I. Stoica: Stateless Core, LNCS 2979, pp. 153-171, 2004.
© Springer-Verlag Berlin Heidelberg 2004

grained monitoring functionality is to minimize the interferences with the system operations. We use two techniques to address this challenge. First, we off-load as much as possible of the processing of log data on an external machine. Second, we use raw IP to send the log data from router's kernel *directly* to the external machine. This way, we avoid context-switching between the kernel and the user level.

To easily configure our system, we have implemented a command line configuration tool. This tool allows us (1) to configure routers as ingress, egress, or core, (2) set-up, modify, and tear-down a reservation, and (3) set-up the monitoring parameters. To minimize the interferences between the configuration operations and data processing, we implement our tool on top of the Internet Control Management Protocol (ICMP). Again, by using ICMP, we avoid context-switching when configuring a router.

The rest of the chapter is organized as follows. The next section presents details of the main operations implemented by the data and control paths. Section 8.2 describes encoding techniques used to efficiently store the DPS state in the packet headers. Section 8.3 presents our packet level monitoring tool, while Section 8.4 describes the configuration tool. Finally, Section 8.5 concludes this chapter.

8.1 Prototype Implementation

For all practical purposes, a FreeBSD PC implements an output queueing router. Output interfaces usually employ a FIFO scheduling discipline, and drop-tail buffer management. Upon a packet arrival, the router performs a routing table lookup, based on the packet destination address, to determine to which output interface the packet should be forwarded. If the output link is idle, the packet is passed directly to the device driver of that interface. Otherwise, the packet is inserted at the tail of the packet queue associated to the interface, by using the IF_ENQUEUE macro.

When the output link becomes idle, or when the network card can accept more packets, a hardware interrupt is generated. This interrupt is processed by the device driver associated with the network card. As a result, the packet at the head of the queue is dequeued (by using the IF_DEQUEUE macro) and sent to the network card.

Our implementation is modular in the sense that it requires few changes to the FreeBSD legacy code. The major change is to replace the queue manipulation operations implemented by IF_ENQUEUE and IF_DEQUEUE. In particular, we replace

```
IF_ENQUEUE(&ifp->if_snd, mb_head);
```

with

```
if (ifp->node_type)
    dpsEnqueue(ifp, (void **)&mb_head);
```

```
else
    IF_ENQUEUE(&ifp->if_snd, mb_head);
```

In other words, we first check whether the router is configured as a DPS router or not, and if yes, we call *dpsEnqueue* to enqueue the packet[1] in a special data structure maintained by DPS. If not, we simply use the default IF_DEQUEUE macro. Similarly, we replace

```
IF_DEQUEUE(&ifp->if_snd, mb_head);
```

with

```
if (ifp->node_type)
    dpsDequeue(ifp, &mb_head);
else
    IF_DEQUEUE(&ifp->if_snd, mb_head);
```

In the current implementation, we support two device drivers: the Intel EtherExpress Pro/100B PCI Fast Ethernet driver, and the DEC 21040 PCI Ethernet Controller. As a result, the above changes are limited to three files: *if_ethersubr.c*, which implements the Ethernet device-dependent functionality, and *if_de.c* and *if_fxp.c* that implement the device-dependent functionalities for the two drivers we support.

In the remainder of this section, we present implementation details for the operations performed by our solution on both the data and control paths. All the processing is encapsulated by the *dpsEnqueue* and *dpsDequeue* functions. The implementation is about 5,000 lines.

8.1.1 Updating State in IP Header

Before forwarding a packet, a core router may update the information carried in its header. For example, in the case of Core-Stateless Fair Queue (CSFQ), if the estimated rate carried by a packet is larger than the link's fair rate, the router has to update the state carried by the packet (see Figure 4.3). Similarly, with Core-Jitter Virtual Clock (CJVC), a router has to insert the "ahead of schedule" variable, g, before forwarding the packet (see Figure 5.3).

Updating the state involves two operations: encoding the state, and updating the IP checksum. State encoding is discussed in details in Section 8.2. The checksum is computed as the 1's complement over the IP header, and therefore it can be easily updated in an incremental fashion.

---

[1]Internally, FreeBSD stores packets in a special data structure called *mbuf*. An ordinary mbuf can store up to 108 bytes of data. If a packet is larger than 108 bytes, a chain of mbufs is used to store the packet. mb_head represents the pointer to the first mbuf in the chain.

8.1.2 Data Path

In this section we present the implementation details of the main operations performed by ingress and core routers on the data path: *packet classification*, *buffer management*, and *packet scheduling*. Since the goal of our solution is to eliminate the per flow state from core routers, we will concentrate on operation complexity at these routers. (The exception is packet classification which is performed by ingress routers only.)

Packet Classification Recall that in SCORE/DPS only ingress routers need to perform packet classification. Core routers do not, as they process packets based only on the information carried in the packet headers.

Our current prototype offers only limited classification capabilities.[2] In particular, a class is defined by *fully* specifying the source and destination IP addresses, the source and destination port numbers, and the protocol type. This allows us to implement packet classification by using a simple *hash table* [26], in which the keys are computed by xor-ing all fields in the IP header that are used for classification.

Buffer Management With both CJVC and CSFQ, routers do not need to maintain any per flow state to perform buffer management. In particular, with CSFQ, packets are dropped probabilistically based on the flow estimated rate carried by the packet header, and the link fair rate (see Figure 4.3 and 4.4). In the case of CJVC, we simply use a single drop-tail queue for all the guaranteed traffic. This is made possible by the fact that, as we have discussed in Section 5.3.3, a relatively small queue can guarantee with a very high probability that no packets are ever dropped. For example, for all practical purposes, a 8,000 packet queue can handle up to one million flows!

Packet Scheduling With CSFQ, the scheduling is trivial as the packets are transmitted on a Fist-In-First-Out basis. In contrast, CJVC, is significantly more complex to implement. With CJVC, each packet is assigned an eligible time and a deadline upon its arrival. A packet becomes eligible when the system time exceeds the packet's eligible time. The eligible packets are then transmitted in the increasing order of their deadlines. Similar to other rate-controlled algorithms [112, 126], we implement CJVC using a combination of a calendar queue and a priority queue [10].

The calendar queue performs rate-regulation, i.e., it stores packets until they become eligible. The calendar queue is organized as a circular queue in which every entry corresponds to a time interval [15]. Each entry stores the list of all packets that have the eligible times in the range assigned to that entry. When the system time advances past an entry, all packets in that

---

[2]This can be easily fixed, by replacing the current classifier with the more complete packet classifier developed in the context of the Darwin project at Carnegie Mellon University [19].

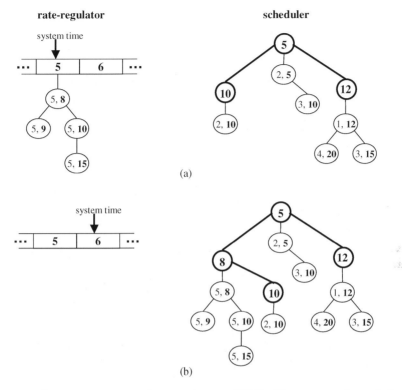

Fig. 8.1. Data structures used to implement CJVC. The rate-regulator is implemented by a calendar queue, while the scheduler is implemented by a two-level priority queue. Each node at the second level (and each node in the calendar queue) represents a packet. The first number represents the packet's eligible time; the second its deadline. (a) and (b) show the data structures before and after the system time advances from 5 to 6. Note that all packets that become eligible are moved in *one* operation to the scheduler data structure.

entry become eligible, and they are moved in a priority queue. The priority queue maintains all eligible packets in decreasing order of their deadlines. Scheduling a new packet reduces then to the selection of the packet at the head of the priority queue, i.e., the packet with the smallest deadline.

The algorithmic complexity of the calendar queue depends on how packets are stored within an entry. If they are maintained in a linked list, then the insertion and deletion of a packet are constant time operations. In contrast, insertion and deletion of a packet in the priority queue takes $O(\log n)$, where n is the number of packets stored in the priority queue [26]. Thus the total cost per packet manipulation is $O(\log n)$.

While the above implementation is quite straightforward, it has a significant drawback. As described above, when the system time advances, *all* packets in an entry become simultaneously eligible, and they have to be transferred in the scheduler priority queue. The problem is that, at least theoretically, the number of packets that become eligible can be unbounded. Moreover, since the packets are maintained in a simply linked list, we need to move them one-by-one! Thus, if we have to move m packets, this will take $O(m \log n)$. Clearly, in a large system this solution is unacceptable, as it may result in packets missing their deadlines due to the large processing time required to move all packets from a calendar queue entry.

To address this problem, we store all packets that belong to the same calendar queue entry in a priority queue ordered by the packets' deadlines. This way, when these packets become eligible, we can move all of them in the scheduler data structure in *one* step. To achieve this, we change the scheduler data structure to a two-level priority queue. Figure 8.1 illustrates the new data structures in the context of a movement operation. Thus, with the new data structures, we reduce the time it takes to move all packets from a calendar queue entry into the scheduler data structure to $O(\log n)$.

A final point worth noting is that the number of packets, n, that are stored at any time in these data structures is, with a very high probability, much smaller than the total number of active flows. As discussed in Section 5.3.3, this is an artifact of the fact that flows are aggressively regulated at each node in the SCORE domain.

8.1.3 Control Path

Providing per flow isolation does not require any operations on the control path. In contrast, providing guaranteed services requires a signaling mechanism to perform per flow admission control. However, as described in Section 5.4, the implementation of admission control at the core routers is quite straightforward. For each output link, a router maintains a value B that is updated every time a new packet arrives. Based on this value, the router periodically updates the upper bound of the aggregate reservation R_{bound} (see algorithm in Figure 5.6). Finally, to decide whether to accept a new reservation or not, a core router just needs to perform a few arithmetic operations (again, see Figure 5.6).

8.2 Carrying State in Data Packets

In order to eliminate the need for maintaining per flow state at each router, our DPS based algorithms require packets to carry state in their headers. Since there is limited space in protocol headers and most header bits have been allocated, the main challenge to implementing these algorithms is to (a) find space in the packet header for storing DPS variables and at the same time remain fully compatible with current standards and protocols; and (b) efficiently encode state variables so that they fit in the available space without introducing too much inaccuracy. In the remainder of the section, we present our approach to address the above two problems in IPv4 networks.

There are at least three alternatives to encode the DPS state in the packet headers:

1. Use a new IP option. (Note that DPS has been already assigned IP option number 23 by the Internet Assignment Number Authority (IANA) [56].) From the protocol point of view, this is arguably the least intrusive approach, as it requires no changes to IPv4 or IPv6 protocol stacks. The downside is that using the IP option can add a significant overhead. This can happen in two ways. First, most of today's IPv4 routers are very inefficient in processing the IP options [109]. Second, the IP option increases the packet length, which can cause packet fragmentation.

2. Introduce a new header between the link layer and the network layer, similar to the way labels are transported in Multi-Protocol Label Switching (MPLS) [17].

    Like the previous approach, this approach does not require any changes in the IPv4/IPv6 protocol stack. However, since each router has to *know* about the existence of the extra header in order to correctly process the IP packets, this approach requires changes in *all* routers, no matter whether this is needed to correctly implement the DPS algorithms or not. In contrast, with the IP option approach, if a router does not understand a new IP option, it will simply ignore it. In practice, this can be an important distinction, as many of today's core routers are typically uncongested. Thus, if we were to implement a service like flow protection, with the IP option approach we don't need to touch these routers, while with this approach we need to change all of them.

    An additional problem is that this approach requires us to devise different solutions for different link layer technologies. Finally, note that it also suffers from a fragmentation problem, since the addition of the extra header will increase the size of the packet.

    Another option to implement this approach would be to leverage MPLS, whenever possible. In particular, in a network that implements MPLS, a solution would be to use an extra MPLS label to encode the DPS state.

3. Insert the DPS state in the IP header. The main advantage of this approach is that it avoids the penalty imposed by most IPv4 routers in processing the IP options, or the need of devising different solutions for different technologies as it would have been required by introducing a new header between the link and network layers. The main problem however is finding enough space to insert the extra information.

While the first two approaches are quite general and can potentially provide large space for encoding state variables, due to performance and easy of implementation reasons, we choose the third approach in our current proto-type.[3]

8.2.1 Carrying State in IP Header

The main challenge to carrying state in the IP header is to find enough space to insert this information in the header while remaining compatible with current standards and protocols. In particular, we want the network domain to be transparent to end-to-end protocols, i.e., the egress node should restore the fields changed by ingress and core nodes to their original values. To achieve this goal, we first use four bits from the type of service (TOS) byte (now renamed the Differentiated Service (DS) field) which are specifically allocated for local and experimental use [75]. In addition, we observe that there is an *ip_off* field of 13 bits in the IPv4 header to support packet fragmentation/reassembly which is rarely used. For example, by analyzing the traces of over 1.7 million packets on an OC-3 link [77], we found that less than 0.22% of all packets were fragments.

Therefore, in most cases it is possible to use *ip_off* field to encode the DPS values. This idea can be implemented as follows. When a packet arrives at an ingress node, the node checks whether a packet is a fragment or needs to be fragmented. If neither of these are true, the *ip_off* field in the packet header will be used to encode DPS values. When the packet reaches the egress node, the *ip_off* is cleared. Otherwise, if the packet is a fragment, it is forwarded as a best-effort packet. In this way, the use of *ip_off* is transparent outside the domain. We believe that forwarding a fragment as a best-effort packet is acceptable in practice, as end-points can easily avoid fragmentation by using a Minimum Transfer Unit (MTU) discovery mechanism. Also note that in the above solution we implicitly assume that packets can be fragmented only by egress nodes.

In summary, we have up to 17 bits available in the current IPv4 header to encode four state variables (see Figure 8.4). The next section discusses some general techniques to efficiently encode the DPS state.

8.2.2 Efficient State Encoding

One simple solution to efficiently encode the state is to restrict each state variable to only a small number of possible values. For example if a state variable is limited to eight values, only three bits are needed to represent it. While this can be a reasonable solution in practice, in our implementation we use a floating point like representation to represent a wider range of values. To further optimize the use of the available space we employ two additional techniques. First, we use the floating point format *only* to represent the *largest*

---

[3]This choice can be also seen as an useful exercise that forces us to aggressively encode the state in the scarce space available in the IP header.

```
void intToFP(int val, int *mantissa, int *exponent) {
  int nbits = get_num_bits(val);
  if (nbits <= m) {
    *mantissa = val;
    *exponent = (1 << n) - 1;
  } else {
    *exponent = nbits - m - 1;
    *mantissa = (val >> *exponent) - (1 << m);
  }
}

int FPToInt(int mantissa, int exponent) {
  int tmp;
  if (exponent == ((1 << n) - 1))
    return mantissa;
  tmp = mantissa | (1 << m);
  return (tmp << exponent)
}
```

Fig. 8.2. The C code for converting between integer and floating point formats. m represents the number of bits used by the mantissa; n represents the number of bits in the exponent. Only positive values are represented. The exponent is computed such that the first bit of the mantissa is always 1, when the number is $\geq 2^m$. By omitting this bit, we gain an extra bit in precision. If the number is $< 2^m$ we set by convention the exponent to $2^n - 1$ to indicate this.

value, and then represent the other value(s) as a fraction of the largest value. In this way we are able to represent a much larger range of possible values. Second, in the case in which there are states which are not required to be simultaneously encoded in the same packet, we use the same field to encode them. Next, we present the floating point like format used to encode large values.

Assume that a is the largest value carried by the packet, where a is a positive integer. To represent a we use an m bit mantissa and an n bit exponent. Since $a \geq 0$, it is possible to gain an extra bit for mantissa. For this we consider two cases: (a) if $a \geq 2^m$ we represent a as the closest value of the form $u2^v$, where $2^m \leq u \leq 2^{m+1}$. Then, since the $m + 1$-th most significant bit in the u's representation is always 1, we can ignore it. As an example, assume $m = 3$, $n = 4$, and $a = 19 = 10011$. Then 19 is represented as $18 = u \times 2^v$, where $u = 9 = 1001$ and $v = 1$. By ignoring the first bit in the representation of u the mantissa will store 001, while the exponent will be 1. (b) On the other hand, if $a < 2^m$, the mantissa will contain a, while the exponent will be $2^n - 1$. For example, for $m = 3$, $n = 4$, and $a = 6 = 110$, the mantissa is 110, while the exponent is 1111. Converting from one format to another can be efficiently implemented. Figure 8.2 shows the conversion code in C. For simplicity, we assume that integers are truncated rather than rounded when represented in floating point.

By using m bits for mantissa and n for exponent, we can represent any integer in the range $[0..(2^{m+1} - 1) \times (2^{2^n - 1})]$ with a relative error bounded by $(-1/2^{m+1}, 1/2^{m+1})$. For example, with 7 bits, by allocating 3 for mantissa and 4 for exponent, we can represent any integer in the range $[1..15 \times 2^{15}]$ with a relative error bounded by $(-6.25\%, 6.25\%)$. Note that these bounds are not necessary tight. Indeed, in this example, the worst cases occur when encoding the numbers 271, and 273, which both have a mantissa of 8. In particular, $271 = 100001111$ is encoded as $u = 1000$, $v = 5$, and has a relative error of $(8 \times 2^5 - 271)/271 = -0.0554 = -5.54\%$, while $273 = 100010001$ is encoded as $u = 1001$, $v = 5$, and has a relative error of 5.55%.

If another value $b \le a$ is carried by the packet we store it as the fraction $f = b/a$. Assuming that we use m_1 bits to represent f, the *absolute* error is bounded by $(-1/(2(2^{m_1} - 1)), 1/(2(2^{m_1} - 1)))$. The -1 in the denominators is a result of mapping 2^{m_1} values to $[0, 1]$, with $2^{m_1} - 1$ representing 1. Finally, it is easy to show that by representing a in floating point format with m bits for mantissa and n bits for exponent, and by using m_1 bits to encode b, the *relative* error of $a + b$, denoted $RelErr(a + b)$, is bounded by

$$-\frac{1}{2^{m+1}} - \frac{1}{2^{m_1+1} - 2} < RelErr(a + b) < \frac{1}{2^{m+1}} + \frac{1}{2^{m_1+1} - 2}, \quad (8.1)$$

where we ignore the second order term $1/(2^{m+1}(2^{m_1+1} - 2))$.

In the next sections, we describe in detail the state encoding in the case of both guaranteed and flow protection services.

8.2.3 State Encoding for Guaranteed Service

With the Guaranteed service, we use three types of packets to carry the DPS state: (1) data packets, (2) dummy packets, and (3) reservation request packets. Below, we describe the state encoding in each of these cases:

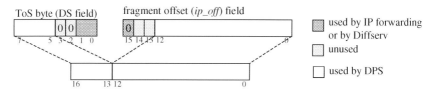

Fig. 8.3. For carrying DPS state we use the four bits from the TOS byte (or DS field) reserved for local use and experimental purposes, and up to 13 bits from the IP fragment offset.

Service type	Packet type	State encoding	Var. Types	State semantics

Guaranteed service	Data packet	`16 15 14 13  11 10   7 6      0` `[1][1][T][F1][F2][F3]`	T – flag (1 bit)
			$F1$ – integer (3 bit)
			$F2$ – integer (4 bit)
			$F3$ – float (3 bit mantissa, 4 bit exponent)
				$d = current\_time + F3 \times (F2+1)$
				if $(T=0)$
				  $e = d - F1 \times F3$
				else
				  $e = current\_time$
				  $b = F1$

	Dummy packet	`16 15 14 13 12 11 10 9    0` `[1][0][1][0][0][0][0][F1]`	$F1$ – float (6 bit mantissa, 4 bit exponent)	$b = 1$
	Rsv. packet	`16 15 14 13 12 11   7 6    0` `[1][0][0][D][R][F1][F2]`	D – flag (1 bit) R – flag (1 bit) $F1$ – integer (5 bit) $F3$ – float (3 bit mantissa, 4 bit exponent)	$h = F1$ $r = F2$
Relative diff. (LIRA)	Data packet	`16 15 14 13 12         0` `[0][1][P][M][F1]`	P – flag (1 bit) M – flag (1 bit) $F1$ – integer (13 bit)	$prefered = P$ $marked = M$ $label = F1$
Flow protection (CSFQ)	Data packet	`16 15 14 13 12 11 10 9    0` `[0][0][0][0][0][0][0][F1]`	$F1$ – float (5 bit mantissa, 5 bit exponent)	$\hat{r} = F1$

Fig. 8.4. State encoding in the case of providing guaranteed service, and flow protection.

Data Packet As described in Chapter 5, there are four pieces of state that need to be encoded in a data packet: (1) the reserved rate r or equivalently l/r, where l is the packet length, (2) the slack delay δ, as computed by Eq. (5.10), (3) the amount of time g by which the packet was transmitted ahead of schedule at the previous node, and (4) b, as computed by Eq. (5.12). The first three pieces of state are used to perform packet scheduling, while the last one is used to perform admission control. All are positive values. Figure 8.4 shows the state encoding in this case. The first two bits represent the code that identifies the state format. The rest of the 15 bits are allocated to four fields:

- a 1-bit flag, called T, that specifies how the next field, $F1$, is interpreted
- a 3-bit field $F1$. If $T = 0$, then $F1$ encodes the b value. Otherwise, it encodes $(l/r)/F3$.
- a 4-bit field $F2$ that encodes $g/F3$
- a 7-bit field $F3$ that encodes $l/r + \delta$.

We make several observations. First, since $F3$ encodes the largest value among all fields, we represent it in floating point format. By using this format, with seven bits we can represent any positive number in the range $[1..15 \times 2^{15}]$, with a relative error within $(-6.25\%, 6.25\%)$. Second, since the deadline determines the delay guarantees, we use a representation that trades eligible time accuracy for deadline accuracy.[4] In particular, the deadline is computed

---

[4] As long as the eligible time value is under-estimated, its inaccuracy will affect only the scheduling complexity, as the packet may become eligible earlier.

as $d = current\_time + F2 * F3 + F3 \simeq current\_time + g + l/r + \delta$. If T is 0, the eligible time is computed as $e = d - F1 * F3 \simeq current\_time + g + \delta$. $F1$ uses only three bits and its value is computed such that $F1 * F3$ always overestimates l/r. If T is 1, the eligible time is computed simply as $e = current\_time$. Third, we express b in 1 KB units. In this way we eliminate the need for each packet to carry the b value. In fact, if a flow sends at its reserved rate, only one packet in every eight packets needs to carry the b value. This observation, combined with the fact that the under-estimation of the packet eligible time does not affect the guaranteed delay of the flow, allows us to alternatively encode either b or $(l/r)/F3$ in $F1$, without impacting the correctness of our algorithms.

Dummy Packet As described in Section 5.4, if a flow does not send any data packet for a period of time larger than T_I, the ingress router has to send a dummy packet. In practice, this can be either a packet newly generated packet by the ingress, or a best-effort packet that happens to traverse the same path. The state in the dummy packet has the code 1010000, and it carries only the b value in the $F1$ field. Note that using a long code leaves extra room for defining other state formats in the future. Dummy packets are forwarded with a priority higher than the priority of any other traffic excepting the guaranteed traffic.

Reservation Request Packet This packet is generated by an ingress router upon receiving a reservation request. The ingress forwards this packet to the corresponding egress router in order to perform the requested reservation. Upon receiving this packet, a core router performs admission control, and updates the state carried by the packet accordingly. When an egress router receives a reservation request packet, it sends it back to the ingress. Upon receiving this packet, the ingress router makes the final decision. Like the dummy packet, the reservation request packet can be either a newly generated packet, or a best-effort packet that happens to traverse the same path.

The code of the state format carried by this packet is 101. The packet carries four fields in its header:

- a 1-bit field, D, that denotes whether the packet traverses the forward or the backward path. If $D = 1$, the packet traverses the forwarding path; otherwise the packet is on its way back to the ingress router that generated it.
- a 1-bit field, A, which specifies the current state of the reservation request. If $A = 1$, then all previous routers have accepted the reservation request; otherwise at least one router has denied it.
- a 5-bit field, $F1$, that stores the number of hops, h, on the forwarding path. The field is incremented by each core router along the path. The number of hops is required to compute the slack variable of each packet at the ingress router (see Eq. (5.10)). Note that the current implementation allows a path with at most 32 hops. While this value does not cover the lengths of all

end-to-end routes observed in today's Internet, this value should be enough for an ISP domain, which is the typical SCORE domain.

- a 7-bit field that stores the requested rate. This rate is stored in floating point format, with a 3-bit mantissa and a 4-bit exponent, and is expressed in 1 Kbps units.

8.2.4 State Encoding for LIRA

With LIRA, we use only one type of packet to carry the DPS state information. The first two bits are always 01, and represent the code that identifies the state format. The rest of the 15 bits are divided into three fields as follows:

- a 1-bit flag, P, called the preferred bit. This bit is set by the application or user and indicates the dropping preference of the packet. This bit is immutable, i.e., it cannot be modified by any router in the network.
- a 1-bit flag, M, called the marking bit. This bit is set by the ingress routers of an ISP and indicates whether the packet is in- or out-of profile. When a preferred packet arrives at an ingress node, the node marks it if the user has not exceeded its profile; otherwise the packet is left unmarked. Whenever there is a congestion, a core router always drops unmarked packets first. Irrespective of the congestion state, core routers never change the marked bit. The reason we use two bits (i.e., the marked and the preferred bits) instead of one is that in an Internet environment with multiple ISPs, even if a packet may be out-of profile in some ISPs on the earlier portion of its path, it may still be in-profile in a subsequent ISP. Having the preference bit that is unchanged by upstream ISPs on the path will allow downstream ISPs to make the correct decision.
- a 13-bit integer that represents the packet label. As discussed in Section 6.3.4, a label is computed as the XOR over the identifiers of all routers along the flow's path. The length of a router's identifier is equal to the label's length, i.e., 13 bit. The packet label is initialized by the ingress router and updated by core routers as the packet is forwarded. Note that there is a non-zero probability that two labels may collide, i.e., that two alternate paths between the same routers have the same label. If router identifiers are uniformly distributed, the probability of collision is $1/2^{13} = 1/8192$. While this probability cannot be neglected in practice, there are two points worth noting. First, this problem can be alleviated by having the ISP choose the router identifiers such that the probability of label collision is minimized. Second, note that even if two alternate paths have the same label, the worst thing that may happen is an alternate path will be ignored. Although this will eventually reduce the link utilization, it will not jeopardize the correctness of our scheme.

8.2.5 State Encoding for CSFQ

As discussed in Chapter 4, with CSFQ, data packets carry only one piece
of state: the estimated rate of the flow. In this case, we define a code of
010000, and a 10-bit field that encodes the estimated rate, by using a floating
point format with a 5-bit mantissa, and a 5-bit exponent. The estimated
rate is initialized by the ingress router and then updated by core routers
whenever the estimated rate is greater than the fair rate on the output link
(see Section 4.3).

By using a long code for the CSFQ state, we leave considerable room for
defining future state formats.

8.2.6 State Encoding Formats for Future Use

By inspecting the state encoding formats in Figure (8.4), it is easy to see
that there is significant room for future extensions. In particular, any state
encoding format that starts with 001, 111, 1011, 0001, 10101, 00001, 101001,
000001, 1010001 or 0000001 is available for future use. Thus, even by restricting ourselves to reusing the space in the IP packet header, we can still
design new DPS based algorithms and mechanisms that (at least at the level
of the state encoding) are fully compatible to the solutions proposed in this
dissertation, and that can use up to 14 bits for specific state encoding.

8.3 System Monitoring

When implementing a network service, a key challenge is discovering whether
the service is working or not, and, more importantly, if it doesn't work, finding
out why. To help answer these questions we need a monitoring tool that can
accurately expose router behavior. Ideally, we want the ability to monitor
router's behavior in *real time*, and at the smallest possible granularity, i.e.,
at the packet level. A key challenge when implementing such a fine grained
monitoring functionality is to minimize the interferences to the monitored
system.

A possible solution that fully addresses this challenge is to use a *hub* or
a similar device before/after each router's interface to replicate the entire
traffic and divert it to an external monitoring machine where it can be processed. Unfortunately, there are two problems with this solution. First, it is
very hard to accurately measure the arrival and the departure times of a
packet. With a simple hub we can infer these times only when packets arrive
at the monitoring machine. However, the packet processing overhead at the
monitoring machine can significantly impact the accuracy of the time estimates. Even with a special device that can insert a timestamp in each packet
when the packet is replicated, the problem is not trivial. This is because the
arrival and departure times of a packet are inserted by two different devices,

so we need a very accurate synchronization clock mechanism to minimize the errors. The second problem with this solution is that it requires an expensive hardware infrastructure.

IP Packet Header

4	IHL	ToS	total length (16 bits)	
identification (16 bits)			flags	fragment offset (13 bits)
TTL		protocol	header checksum (16 bits)	
source IP address (32 bits)				
destination IP address (32 bits)				

+

arrival time (32 bits)
departure time (32 bits)
log type

Fig. 8.5. The information logged for each packet that is monitored. In addition to most fields of the IP header, the router records the arrival and departure times of the packet (i.e., the ones that are shaded). The type of event is also recorded, e.g., packet departure, or packet drop. The DPS state is carried in the ToS and the fragment offset fields.

For these reasons, in our implementation we use an alternate approach. In particular, we instrument the kernel to perform packet level monitoring. For each packet, we record the arrival and the departure times with very high accuracy using the Pentium clock counter. In addition to these times, with each packet we log most of the fields of the IP header, including the DPS state carried by the packet, and a field, called log type, which specifies whether the packet was transmitted or dropped. To minimize the monitoring overhead, we use the ip_output function call to send this information directly from kernel to an external monitoring machine. Using ip_output, instead of ioctl, for transferring log data, avoids unnecessary context switching between the kernel and the user level. In addition, we off-load as much as possible of the log data processing to the monitoring machine. The functions performed by a router are kept to minimum: a router has only to (1) log packet arrival and departure times, and (2) send the log data to the monitoring machine. In turn, the monitoring machine performs the bulk of data processing, such as packet classification, and computing the rate and/or delay of a flow. To further reduce the monitoring overhead, we provide the ability to sample the data traffic. In particular, we can configure the router to probabilistically log one out of every 2, 4, 16, or 64 packets. Finally, to eliminate possible interferences between the data traffic and transferring the log data, we use a different network interface to send out the log information.

To visualize the monitoring information, we have developed a Graphical User Interface (GUI) written in Java. This tool offers the possibility of simultaneously plotting up to nine graphs in real time. Each of the graphs can be independently configured. In particular, the monitoring tool allows us to specify for each graph:

- the network interface to be monitored. This is specified by the IP address of the router to which the interface is attached, and the IP address of the neighboring router that is connected to the interface.
- the flows to be monitored at the selected network interface. In our case, a flow is identified by its source and destination IP addresses, its source and destination port numbers, and the protocol type.
- the parameters to monitor. Currently the tool can plot the throughput and the delay of a flow. The throughput is averaged over a specified time period. The delay can be computed either as the minimum, maximum, or average packet delay over a specified period of time.

This flexibility allows us, for example, to simultaneously visualize the throughput and per-hop delay of a flow at different hops, to visualize the throughput and the delay of a flow at the same hop, or to visualize the throughput of a flow at multiple time scales. As an example, Figure 3.5 shows a screen snapshot of our monitoring tool. The top two plot show the throughputs of three flows at two consecutive hops. The bottom two plots show the delays of the same flows at the same hops.

8.4 System Configuration

As with any software system, we need the ability to configure our system. In particular, we want the ability to configure a router as an ingress, egress, or core, and to set-up the various parameters of the packet classifier and packet scheduler. In addition, in the case of the guaranteed service, we want the ability to set-up, modify, and tear-down a flow reservation. One possible solution would be to use an `ioctl` function call. Unfortunately, this approach has a significant drawback. Since the execution of `ioctl` is quite expensive – it requires context-switching between the kernel and the user level – it can seriously interfere with the processing of the data traffic. At the limit, in the case of the guaranteed traffic, this can result in packets missing their deadlines.

To avoid this problem, we use the Internet Control Management (ICMP) protocol. In particular, we have written a simple command line utility, called `config_dps`, that can be used to configure routers and initialize their parameters via ICMP. To achieve this, we have extended the ICMP protocol by adding two new messages: `ICMP_DPSREQ`, and `ICMP_DPSREPLY`.

8.4.1 Router Configuration

So far we have classified routers as edge and core routers, and the edge routers as ingress and egress routers. While at the conceptual level this classification makes sense, in practice, this distinction is less clear. For example, an edge router can be either an ingress or egress depending on the direction of the traffic that traverses it. As a result, we need to configure at a finer granularity, i.e., at the network interface level, instead of at the router level. We identify an interface by two IP addresses: the address of the router to which the interface is attached, and the address of the downstream router connected to the interface. We then use the following command to configure a network interface:

```
config_dps 2 node next_node type
```

The first parameter of the command, the number 2, specifies that an interface is being configured. The next two parameters specify the interface: node represents the IP address of the router whose interface is being configured, and next_node specifies the IP address of the downstream router that is connected to the interface. Finally type specifies whether this interface is configured as belonging to an ingress, egress, or core router.

8.4.2 Flow Reservation

In the case of the guaranteed service we need the ability to set-up, modify, and tear-down a flow reservation.

To set-up a new reservation we use the following command:

```
config_dps 3 ingress egress src_addr dst_addr src_port dst_port
proto rsv_rate q_size
```

The first parameter, the number 3, specifies that this command requests a new reservation. The next two parameters specify the ingress and egress routers which are the end-points of our reservation. In addition, we use these parameters to identify the interface of the ingress where the flow state is to be instantiated. The next five parameters, src_addr, dst_addr, src_port, dst_port, and proto, respectively, identify the flow. Finally, the last two parameters specify the requested reservation (in Kbps), and the buffer size to be allocated at the ingress for this flow (in packets).

To terminate a reservation, we use the following command:

```
config_dps 4 ingress egress src_addr dst_addr src_port dst_port
```

The first parameter, the number 4, represents the code of the deletion operation. The other parameters have the same meaning as in the previous command.

Finally, the command to update a reservation is virtually identical to setting-up a reservation. In fact, this operation is implemented by tearing down the existing reservation, and creating a new one.

8.4.3 Monitoring

As described in Section 8.3, our implementation provides support for fine grained monitoring. To start monitoring the traffic at a network interface, we use the following command:

```
config_dps 0 node next_node mon_node mon_port level
```

Again, the first parameter, the number 0 in this case, represents the operation code. The next two parameters are used to identify the interface that we want to monitor, while mon_node and mon_port represent the IP address (of the monitoring machine) and the port number where the log data is to be transmitted. Finally, level specifies the sampling level. The routers probabilistically log one packet out of 2^{level} consecutive packets that arrive at the interface. To stop monitoring the traffic at an interface, we use the same command, but with the second parameter set to 1, instead of 0.

8.5 Summary

In this chapter, we have presented a prototype implementation of our solution to provide guaranteed and flow protection services in an IPv4 network. To the best of our knowledge this is the first implementation that provides these services in a core stateless network architecture. We have described the operations performed by both edge and core routers on data and control paths, as well as an approach that allows to efficiently encode the DPS state in the IP header by re-using some of the existing fields. In addition, we have described two support tools that allow packet level monitoring in real-time, and easy system configuration.

We have implemented our solution in FreeBSD 2.2.6, and deployed it in a local test-bed that consists of four PC routers and up to 16 hosts. The results presented in Sections 3.2 and 5.5 show that (1) the overhead introduced by our solution is acceptable, i.e., we can still saturate the 100 Mbps, (2) the overhead increases very little with the number of flows (at least when the number of flows is no larger than 100), and (3) the scheduling mechanisms protect the guaranteed traffic so that its performance is not affected by the best-effort traffic.

While implementing this prototype we have learned two valuable lessons. First, the monitoring tool proved very useful not only in debugging the system, but also in promptly finding unexpected causes of experiment failures. For example, in more than one instance, we were able to easily discover that we do not achieve the expected end-to-end performance, simply because the source does not correctly generate the traffic due to unexpected interference (such as someone inadvertently using the same machine). Second, the fact that our solution does not require the maintenance of distributed per flow state has significantly simplified our implementation. In particular, we were able to implement the entire functionality of the control path, which is notoriously difficult to implement and debug in the case of the stateful solutions such as RSVP [128], in just a couple of days.

However, our implementation suffers from some limitations that we hope to address in the future:

- The current test-bed is a local area network. We would like to perform similar tests in a larger internetwork environment such as CAIRN [16].
- All the experiments were performed with synthetic loads. In the future, we would like to experiment with real applications such as video conferencing or distributed simulations.
- The test-bed consists of only PC workstations. We would like to implement our algorithms in commercial routers.
- The current configuration tool does not offer any protection. Anyone who knows the address of our routers and the command format, can send ICMP messages to re-configure the routers. Providing an encryption mechanism to avoid malicious router re-configuration would be useful.

9 Conclusions and Future Work

In this chapter, we conclude the dissertation by (1) summarizing our contributions, (2) exposing some fundamental limitations of current solutions, and (3) proposing several directions for future work.

9.1 Contributions

One of the most important reasons behind the overwhelming success of the Internet is the stateless nature of its architecture. The fact that routers do not need to maintain any fine grained information about traffic makes the Internet both *scalable* and *robust*. However, these advantages come at a price: today's Internet provides only a minimalist service, the best effort datagram delivery. As the Internet evolves into a global communication infrastructure that is expected to support a plethora of new applications such as IP telephony, interactive TV, and e-commerce, the existing best effort service will no longer be sufficient. As a result, there is an urgent need to provide more powerful services such as guaranteed services and flow protection.

Over the past decade, there has been intense research toward achieving this goal. Two classes of solutions have been proposed: those maintaining the *stateless* property of the original Internet (e.g., Differentiated Services), and those requiring a new *stateful* architecture (e.g., Integrated Services). While stateful solutions can provide more powerful and flexible services such as per flow guaranteed services, and can achieve higher resource utilization, they are less scalable than stateless solutions. In particular, stateful solutions require each router to maintain and manage per flow state on the control path, and to perform per flow classification, scheduling, and buffer management on the data path. Since there can be a large number of active flows in the Internet, it is difficult, if not impossible, to implement such solutions in a scalable fashion. On the other hand, while stateless solutions are much more scalable, they offer weaker services.

The main contribution of this dissertation is to *bridge the long-standing gap between stateful and stateless solutions*. To achieve this goal, we have described a novel technique called Dynamic Packet State (DPS). The key idea behind DPS is that, instead of having routers maintain per flow state, packets carry the state. In this way, routers are still able to process packets

I. Stoica: Stateless Core, LNCS 2979, pp. 173-183, 2004.

on a per flow basis, despite the fact that they do not maintain per flow state. Based on DPS, we have proposed a network architecture called Stateless Core (SCORE) in which core routers do not maintain any per flow state. Yet, by using DPS, we have demonstrated that, in a SCORE network, it is possible to provide services which are as powerful and flexible as the services provided by a stateful network. In particular, we have developed complete solutions to address some of the most important problems in today's Internet:

– **Flow protection** (Chapter 4) We have proposed the first solution to provide protection on a per flow basis without requiring core routers to maintain any per flow state. To achieve this goal, we have used DPS to approximate the functionality of a reference network in which every router implements the Fair Queueing [31] discipline with a SCORE network in which every router implements a novel algorithm, called Core-Stateless Fair Queueing (CSFQ).
– **Guaranteed services** (Chapter 5) We have developed the first solution to provide per flow delay and bandwidth guarantees. We have achieved this goal by using DPS to emulate the functionality of a stateful reference network in which each router implements Jitter Virtual Clock [126] on the data path, and per flow admission control on the control path in a SCORE network. To this end, we have proposed a novel scheduling algorithm, called Core-Jitter Virtual Clock (CJVC), that provides the same end-to-end delay bounds as Jitter Virtual Clock, but, unlike Jitter Virtual Clock, does not require routers to maintain per flow state.
– **Large spatial granularity service** (Chapter 6) We have developed a stateless solution that allows us to provide relative differentiation between traffic aggregates over large numbers of destinations. The most complex mechanism required to implement this service is route-pinning, for which traditional solutions require routers to either maintain per flow state, or maintain state that is proportional to the square of the number of edge routers. By using DPS, we are able to significantly reduce this complexity. In particular, we propose a route-pinning mechanism that requires routers to maintain state which is proportional only to the number of egress routers.

While the above solutions have many scalability and robustness advantages over existing stateful solutions, they still suffer from robustness and scalability limitations in comparison. System robustness is limited by the possibility that a single edge or core router may malfunction, inserting erroneous information in the packet headers, severely impacting performance of the entire network. In Chapter 7, we propose an approach, called "verify-and-protect", that overcomes these limitations. We achieve scalability by pushing the complexity all the way to the end-hosts, thus eliminating the distinction between edge and core routers. To address the trust and robustness issues, all routers statistically verify that the incoming packets are correctly marked.

This approach enables routers to discover and isolate misbehaving end-hosts and routers.

To demonstrate the compatibility of our solutions with the existing protocols, we have presented the design and prototype implementation of the guaranteed service in IPv4 networks. In Chapter 8 we propose both efficient state encoding algorithms, as well as an encoding format for the proposed solutions.

The SCORE/DPS ideas have already made an impact in both research and industrial communities. Since we have published the first papers [101, 104], several new and interesting results have been reported. They include both extensions and improvements to the original CSFQ algorithm [18, 25, 111], and generalizations of our solution to provide guaranteed services [129, 130]. In addition, DPS-like techniques have been used to develop new types of applications such as IP traceback [91]. Furthermore, we observe that it is possible to extend the current Differentiated Service framework [32] to accommodate algorithms using Dynamic Packet State. The key extension needed is to augment each Per Hop Behavior (PHB) with additional space in the packet header for storing PHB specific Dynamic Packet State [107]. Such a paradigm will significantly increase the flexibility and capabilities of the services that can be built with a Diffserv-like architecture.

9.2 Limitations

While in this thesis we have shown that by using the DPS technique it is possible to implement some of the most representative Internet services (for which previous solutions required stateful networks) in a SCORE network, one important question still remains: *What are the limitations of SCORE/DPS based solutions?* More precisely, is there any service implemented by stateful networks that cannot be implemented by a SCORE network? In this section, we informally answer these questions.

In a stateful router, each flow is associated a set of state variables such as the length of a flow's queue and the deadline of the packet at the head of the flow's queue. In addition, a router maintains some global state variables such as buffer occupancy or utilization on the output link. A router processes a packet based on both the per flow state and the global state stored at the router. As an example, upon a packet arrival, a router checks whether the buffer is full, and if not, discards the packet at the tail of the longest queue to make room for the new packet.

In contrast, in SCORE, a core router processes packets based on the state carried in packet headers, instead of per flow state (as these routers do not maintain any such state). Thus, in order to emulate the functionality of a stateful router, a stateless router has to reconstruct the per flow state from the state carried in the packet headers. The question of what are the limitations of SCORE/DPS based solutions reduces then to the question of what types of per flow state cannot be exactly reconstructed by core routers. We are aware of two instances in which current techniques cannot reconstruct the per flow state accurately:

1. *The state of a flow depends on the behavior of other competing flows.* Intuitively, this is because it is very hard, if not impossible, to encode the effects of this dependence in the packet headers. Indeed, this would require a router to know the *future* behavior of the competing flows at the *next* router *before* updating the state in the packet headers.

   Consider the problem of exactly emulating the Fair Queueing (FQ) discipline. Recall that FQ is a packet level realization of a bit-by-bit round robin: if there are n backlogged flows, FQ allocates $1/n$ of the link capacity to each flow. In the packet system, a flow is said to be backlogged if it has at least one packet in the queue.

   The challenge of implementing FQ in a stateless router is that the number of backlogged flows, and therefore the service received by a flow, is a highly dynamic parameter, i.e., it can change every time a new packet arrives or departs. Unfortunately, it is very hard if not impossible for a router to accurately update the number of backlogged flows when such an event occurs. To illustrate this point, consider a packet arrival event. When a packet of an idle flow arrives, the flow becomes backlogged, and the number of backlogged flows increases by one. Because the router does not maintain per flow state, the only way it can infer this information is from the state carried in the packet header (as this is the only new information that enters the system). However, inserting the correct information in the packet header would require up-stream routers to know how many packets are in the flow's queue when the packet arrives at the *current* node. In turn, this would require knowledge of how much service the flow has received at the current node so far, as this determines how many packets are still left in the queue. Unfortunately, even if we were using a feedback protocol to continuously inform the up-stream routers of the state of the current router, the propagation delay poses fundamental limitations on the accuracy of this information. For example, during the time it takes the information to reach the up-stream routers, an arbitrary number of flows may become backlogged!

2. *The state of a flow depends on parameters that are unique to each router.* Intuitively, this is because it is very hard to reconstruct these parameters at each router given the limited space in the packet headers.

   Consider a router that implements the Weighted Fair Queueing (WFQ) scheduling discipline. Similar to FQ, WFQ is a realization of a weighted bit-by-bit round robin: a flow i with weight w_i receives $(w_i / \sum_{j \in B} w_j)$ of the link capacity, where B is the set of backlogged flows. Assume that the flow has a different weight at every router along its path. Then, each packet has to encode all these weights in its header. Unfortunately, in the worst case, encoding these weights requires an amount of state proportional to the length of the path, which is not acceptable in practice.

Not surprisingly, these limitations are reflected in our solutions. The first limitation is the main reason why Core Stateless Fair Queueing is only able to approximate, not emulate, the Fair Queueing discipline. Similarly, it is the main reason why our per flow admission control solution uses an upper bound of the aggregate reservation, instead of the actual aggregate reservation (see Section 5.4). Finally, our decision to use a non-work conserving discipline such as Core-Jitter Virtual Clock to implement guaranteed services is precisely because of this limitation. In particular, the fact that the service received by a flow under a non-work conserving discipline is not affected by the behavior of the competing flows, allows us to broke the dependence between the flow state and the behavior of the competing traffic. This makes it possible to compute the eligible times and the deadlines of a packet at all core routers, as soon as the packet arrives at the ingress router. The potential downside of a non-work conserving discipline is the inability of a flow to use additional resources made available by inactive flows.

As a result of the second limitation, we only consider the cases in which a flow has the same reserved bandwidth or weight at all routers along its path through a SCORE domain. In the case of providing guaranteed services, this restriction can lead to lower resource utilization, as it is difficult to efficiently match the available resources at each router with the flow requirements.

In summary, as a result of these limitations, our SCORE/DPS based solutions cannot exactly match the performance of traditional stateful solutions. In spite of this, as we have demonstrated by analysis and experimental results, our solutions are powerful enough to fully implement some of the most popular per flow network services.

9.3 Future Work

In the next sections, we identify several research directions for future work. Whenever possible, we try to emphasize the main difficulties and possible solutions to address the proposed problems.

9.3.1 Decoupling Bandwidth and Delay Allocations

As described in Chapter 5, our solution to provide guaranteed services associates a *single* parameter to each flow: the flow's reserved rate. While we can provide both per flow delay and bandwidth guarantees by appropriately setting the flow reserved rates, the fact that we are restricted to only one parameter may lead to inefficient resource utilization.

To illustrate this point, consider a flow that sends traffic at a constant rate r, and has fixed size packets of length l. In this case, the worst case delay experienced by a packet at one router is about[1] l/R, where R is the flow's bandwidth reservation. Intuitively, this is because in an idealized model in which the flow traverses only dedicated links of capacity R, it takes a router exactly l/R time to transmit a packet of size l. Assume that the flow requests a per hop delay no larger than D. To meet this requirement, the router has to allocate a bandwidth R, such that $l/R \leq D$, or alternatively $R \geq l/D$. In addition, R should be no smaller than the arrival rate of the flow r. As a result, for a flow with the arrival rate of r and a per hop delay bound of D, a router has to allocate a bandwidth of at least $R = \max(l/D, r)$. Consider a 64 Kbps audio flow that uses 1280 bit packets, and has a per hop delay budget of $D = 5$ ms. To meet this delay requirement, the flow needs to reserve at least $R = l/D = 1280\,\mathrm{bit}/5\,\mathrm{ms} = 256$ Kbps, which is four times more than the flow's rate! Thus, using only one parameter can result in serious resource underutilization.

In the stateful world, several solutions have been proposed to address this problem [90, 106, 124]. A future direction would be to emulate these solutions in the SCORE/DPS framework. The problem is that current solutions to decouple bandwidth and delay allocations use at least two parameters to specify a flow reservation. This significantly complicates both the data and the control path implementations. For example, admission control requires checking whether a two-piece linear function representing the new reservation ever exceeds an n-piece linear function representing the available link resources (capacity), where, in the worst case, n represents the number of flows [90]. Thus, storing the representation of the available capacity requires an amount of state proportional to the number of flows, which is unacceptable for a stateless solution. A possible approach to alleviate this problem would be to restrict the values taken by the parameters that characterize flow reservations. The challenge is doing this without compromising the flexibility offered by decoupling the bandwidth and delay allocations.

9.3.2 Excess Bandwidth Allocation

Our solution to provide guaranteed services is based on a non-work conserving scheduling algorithm, i.e., CJVC. As a result, even if the network is completely idle, a guaranteed flow will receive no more than its reserved rate. While this service is appropriate for many applications such as IP telephony, other applications such as video streaming would prefer a more flexible service in which they can opportunistically take advantage of the unused bandwidth to achieve better quality. In the domain of stateful solutions, there are several algorithms including variants of Weighted Fair Queueing [10, 48, 79], and

---

[1]More precisely the worst case delay is $l/R + l_{max}/C$, where l_{max} represents the maximum length of any packet that traverses the link, and C represents the link capacity. The term l_{max}/C accounts for the fact that the packet transmission is not preemptive. However, since in general $C \gg r$, we ignore this term here.

Fair Service Curve [106] that provide the ability to share the excess (unused) bandwidth.

In this context, it would be interesting to develop stateless algorithms that are able to achieve excess bandwidth sharing, while still providing guaranteed services. As discussed in Section 9.2, the main problem is that it is very hard for DPS algorithms such as CJVC to adapt to very rapid changes of excess bandwidth available at core routers. In the case of CJVC this is because the scheduling parameters are computed when the packet arrives at the ingress router, a point at which it is very hard if not impossible to accurately predict what *will* be the excess bandwidth when the packet arrives at a particular core router. There are at least two general approaches to alleviate this problem: (1) use a feedback mechanism similar to LIRA to inform egress routers about the excess bandwidth available at core routers, and (2) have core routers compute packet scheduling parameters based on both the state carried by the packet headers and some internal state maintained by the router. The main challenge of the first approach is to balance the freshness of the information maintained at ingress routers regarding the excess bandwidth inside the network with the overhead of the feedback mechanism. The main challenge of the second approach is to maintain the bandwidth and delay guarantees without increasing the scheduling complexity. This is hard because the complexity of algorithms such as CJVC depend directly on the buffer size (see Section 5.3.3), and the buffer size at core routers will significantly increase as a result of allowing flows to use excess bandwidth [79].

9.3.3 Link Sharing

While most of the previous research directed at providing better services in packet switching networks have focused on providing guaranteed services or protection for each individual flow, several recent works [11, 39, 92] have argued that it is also important to support hierarchical link-sharing service.

In hierarchical link-sharing, there is a class hierarchy associated with each link that specifies the resource allocation policy for the link. A class represents a traffic stream or some aggregate of traffic streams that are grouped according to administrative affiliation, protocol, traffic type, or other criteria. Figure 9.1 shows an example class hierarchy for a 45 Mbps link that is shared by two organizations, Carnegie Mellon University (CMU) and University of Pittsburgh (U. Pitt). Below each of the two organization classes, there are classes grouped based on traffic types. Each class is associated with its resource requirements, in this case, a bandwidth, which is the minimum amount of service that the traffic of the class should receive when there is enough demand.

There are several important goals that the hierarchical link-sharing service aims to achieve. First, each class should receive a certain minimum amount of resource if there is enough demand. In the example, CMU's traffic should receive at least 25 Mbps of bandwidth during a period when the aggregate

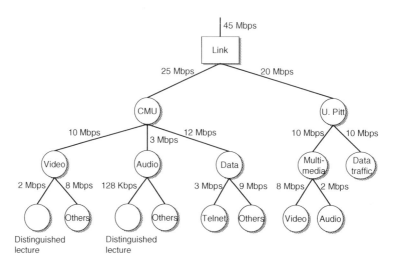

Fig. 9.1. An example of Link-Sharing Hierarchy.

traffic from CMU has a higher arrival rate. Second, at each level of the hierarchy, active children should be able to use the excess bandwidth made available by the inactive children. In the case when there is no audio or video traffic from CMU, the data traffic from CMU should be able to use all the bandwidth allocated to CMU (25 Mbps). Finally, we should be able to provide both bandwidth and delay guarantees to leaf classes, eventually by decoupling the bandwidth and delay allocations. In the example, the CMU Distinguished Lecture video and audio classes are two leaf classes that require both bandwidth and delay guarantees.

In short, hierarchical link-sharing aims to provide (1) bandwidth and delay guarantees at leaf classes, (2) bandwidth guarantees at interior classes, and (3) excess bandwidth distribution among children classes. Due to the service complexity, it should come as no surprise that all current solutions require routers to maintain per class state [11, 39, 92]. A natural research direction would be to implement the link-sharing service in a SCORE network. Providing such a service is challenging because we have to deal with both limitations discussed in Section 9.2: the first limitation makes it difficult to provide excess bandwidth distribution; the second limitation makes it difficult to encode the reservation and the position of each ancestor class the packet belongs to at each router along its paths. A possible solution would be to come up with "reasonable" restrictions that allow efficient state encoding without significantly compromising the flexibility and the utilization offered by existing stateful solutions.

9.3.4 Multicast

The solutions presented in this dissertation were primarily designed for unicast traffic. An interesting direction for future work would be to extend them for multicast. In the case of CSFQ this is straightforward: since packet replication does not affect the flow rate along an individual link, when a router replicates a packet, it just needs to copy the DPS state into the replica headers. In contrast, extending our guaranteed service solution to support multicast would be more difficult. Part of the challenge would be to come up with acceptable service semantics. For example, do we want to delay all packets by the same amount no matter what paths they traverse, or do we want to achieve the minimum delay on each individual path? Do we want to allocate the minimum bandwidth that is available to all receivers, or do we want the ability to allocate different bandwidths on different paths? One important observation that simplifies the problem is that in all traditional multicast solutions, at least the branching routers in the multicast tree have to maintain per group state. By leveraging this state, it would be possible to update the DPS state carried by the packets as a function of the branch they follow. In theory, this will allow us to provide service differentiation on a per path basis.

Another future direction in the context of multicast would be to use DPS to implement the multicast service itself. A straightforward approach would be for the sender to insert the list of all receivers' IP addresses in the packet headers. This would eliminate the need for routers to keep any multicast state: when a packet arrives, a router inspects the list of the addresses in the packet header and replicates the packet to each output port that corresponds to an address in the list. While simple, this solution has a fundamental limitation: the state carried by the packets increases with the number of receivers. In a group with hundreds of receivers, we can easily reach the situation when we have to transmit more DPS state than data information! Note that this problem is an instance of the second limitation discussed in Section 9.2, i.e., per group (flow) routing state maintained by traditional multicast schemes is unique for each router. A possible solution would be to find a more efficient encoding of the forwarding state, and, eventually, partition it between the packet headers and core routers.

9.3.5 Verifiable End-to-End Protocols

As described in Chapter 7, the "verify-and-protect" approach can overcome some of the robustness and scalability limitations of the SCORE/DPS framework. We believe, however, that this is a far more general and powerful approach that can be used to design new network services and protocols. Usually, whenever we implement a network service without support from the network, we make the (implicit) assumption that users *cooperate*. For example, the recently proposed endpoint admission control algorithms assume that (1) each user probes the network to detect the level of congestion, and then (2) it sends traffic only *if* the level of congestion is sufficiently low [14]. Unfortunately, in an economic environment like today's Internet, there is no strong incentive for users to cooperate. For example, a user may choose to send traffic, even if the congestion level is high, in the hope that it will force other users to give up and release their bandwidth. A natural way to create the incentive for users to cooperate is to punish them if they don't. However, this requires the ability to *identify* a malicious user, i.e., the ability to *verify* its behavior. We believe that verifiability should be a *key property* of any end-to-end protocol, and not an afterthought as it happens today. In this context, we believe that designing protocols and algorithms with verifiable behaviors is a very important topic for future work.

9.3.6 Incremental Deployability

SCORE/DPS solutions described in this dissertation require changes to *all* routers within a network domain. This is a serious limitation that may delay or even preclude the deployment of SCORE/DPS solutions in the Internet. An important direction for future work is to alleviate or remove, if possible, this limitation.

One approach would be to develop solutions that require only a *subset* of routers to be changed. One example would be to study what levels of bandwidth and delay guarantees can be provided by a network in which only the edge nodes are changed. The key difficulty would be to coordinate the actions of these routers. Furthermore, this coordination would need to happen at a very small time scale, because of the rapid changes in traffic characteristics. DPS represents an ideal starting point to the development of such mechanisms, as it allows the exchange of traffic information at the smallest possible granularity: on a per packet basis.

Another approach would be to build an overlay network consisting of high performance sites, called Point-of-Presence's (PoPs), connected by high quality virtual links. An example would be to provide per flow bandwidth allocation by using the Premium service to guarantee capacity along each virtual link, and a CJVC like algorithm to manage this capacity among the competing flows. Another example would be to monitor the available bandwidth along each virtual link, and use this information together with a CSFQ-like algorithm to provide fair bandwidth allocation across the overlay network. One possible approach to improve the quality of the virtual links would be to construct the overlay such that each virtual link traverses no more than one ISP. There are two reasons for this restriction. First, since the traffic between any two neighbor PoPs traverses only one IPS, and since an ISP has generally full control of its resources, it would be much easier to provide a strong semantic service. Second, it would be much easier to verify, and therefore to enforce, the service agreement between the neighbor PoPs and the ISP that handles their traffic. If one of the two PoPs detects that the service agreement is broken, then it can conclude that this is because the ISP that carries the traffic does not honor its agreement. In contrast, had the traffic between the two PoPs traverse more that one ISP, it would have been very hard to identify which ISP was to blame.

9.3.7 General Framework

In this dissertation we have demonstrated by examples that it is possible to provide network services with per flow semantics in a stateless network architecture. A very interesting theoretical question is: *what is the class of algorithms and services that can be emulated or approximated in the SCORE/DPS framework?* In Section 9.2 we informally discuss two of the limitations of the current solutions. The next step would be to develop a theoretical framework that precisely formulates the limitations and answers the previous question. Such a framework would provide us with a much better understanding of what we can and what we cannot do in the SCORE/DPS framework. A related question of practical interest is: *can we come up with a general methodology that allows us to transform a stateful network into a stateless network while preserving its functionality?*

9.4 Final Remarks

In this dissertation, we have presented the first solution that can provide services as powerful and as flexible as the ones implemented by a stateful network using a stateless network. To illustrate the power and the generality of our solution, we have implemented three of the most important services proposed in the context of today's Internet: providing guaranteed services, differentiated services, and flow protection. While it is hard to predict the exact course of research in this area, we believe that the door has been opened to many new and challenging problems of great practical importance and theoretical interest.

A Performance Bounds for CSFQ

In this appendix we give the proof of Theorem 1 (Section 4.3.4). Recall that for simplicity we make the following assumptions: (1) the fair share α is fixed, (2) there is no buffering and therefore the drop probability is given by Eq. (4.2), and (3) when a packet arrives, a fraction of that packet equal to the flow's forwarding probability is transmitted. The proof is based on two intermediate results given in Lemmas 1 and 2, respectively. Lemma 1 gives the upper bound for the excess service received by a flow with weight w during an arbitrary time period in which the estimated rate of the flow, as given by Eq. (4.3), does *not* exceed the flow's fair rate, i.e., $w\alpha$. Similarly, Lemma 2 gives the upper bound for the excess service received by a flow during an arbitrary period in which the flow's estimated rate is never *smaller* than its fair rate.

First, we give two well known inequalities that are subsequently used in the proofs:

$$\frac{x}{1 - e^{-x}} > 1, \quad x > 0, \tag{A.1}$$

$$\frac{xe^{-x}}{1 - e^{-x}} < 1, \quad x > 0. \tag{A.2}$$

Lemma 1 *Consider a link with a normalized fair rate α, and a flow with weight w. The excess service received by the flow during any interval $I = [t', t'')$, when its estimated rate r does not exceed its fair rate $r_\alpha = w\alpha$, i.e., $r(t) \le w\alpha$, $\forall t \in I$, is bounded above by*

$$r_\alpha K + l_{max}, \tag{A.3}$$

where l_{max} represents the maximum length of a packet.

Proof. Without loss of generality assume that exactly n packets of the flow are received during the interval I. Let $t_i \in I$, $(1 \le i \le n)$ be the arrival time of the i-th packet, and let l_i denote its length. According to Eq (4.3), we have

$$r_i = (1 - e^{-T_i/K})\frac{l_i}{T_i} + e^{-T_i/K}r_{i-1}, \quad 1 \le i \le n, \tag{A.4}$$

where $T_i = t_i - t_{i-1}$, and r_0 represents the initial estimated rate. If $r_0 > 0$, t_0 is assumed to be the time when the last packet was received before t'. Otherwise, if $r_0 = 0$ then we take $t_0 = -\infty$.

Since by hypothesis $r_i \leq r_\alpha$ $(1 \leq i \leq n)$ it follows that *all* packets are forwarded and therefore the total number of bits sent during the interval I is $\sum_{i=1}^{n} l_i$. Thus, our problem can be reformulated to be

$$\max \left(\sum_{i=1}^{n} l_i \right),$$
(A.5)

subject to

$$r_i \leq r_\alpha, \quad 1 \leq i \leq n.$$
(A.6)

From Eq. (A.4) it follows that

$$l_i = \frac{r_i - r_{i-1} e^{-T_i/K}}{1 - e^{-T_i/K}} T_i, \quad 2 \leq i \leq n.$$
(A.7)

The above equation does not apply for $i = 1$, since it is not well defined for the case in which $r_0 = 0$. Recall that in this case we take $t_0 = -\infty$, and therefore $T_1 = \infty$. Further, define

$$F(r_1, r_2, \ldots, r_n) = \sum_{i=2}^{n} l_i.$$
(A.8)

Our goal then is to maximize $F(r_1, r_2, \ldots, r_n)$. By plugging l_i from Eq. (A.7) into $F(r_1, r_2, \ldots, r_n)$ and taking the derivate with respect to r_i $(2 \leq i \leq n)$ we obtain

$$\frac{\partial F(r_1, r_2, \ldots, r_n)}{\partial r_i} = \frac{T_i}{1 - e^{-T_i/K}} - \frac{T_{i+1} e^{-T_{i+1}/K}}{1 - e^{-T_{i+1}/K}}, \quad 2 \leq i < n. \quad \text{(A.9)}$$

and

$$\frac{\partial F(r_1, r_2, \ldots, r_n)}{\partial r_n} = \frac{T_n}{1 - e^{-T_n/K}}.$$
(A.10)

By using Eqs. (A.1) and (A.2) (and after making substitutions $T_i \to x_1 K$, $T_{i+1} \to x_2 K$, and $T_n \to x_3 K$) we have

$$\frac{\partial F(r_1, r_2, \ldots, r_n)}{\partial r_i} > 0, \quad 2 \leq i \leq n.$$
(A.11)

Thus, $F(r_1, r_2, \cdots, r_n)$ is maximized when $r_2, r_3, \ldots, r_n$ achieve their maximum value, which in our case is r_α. Consequently, we have

$$F(r_1, r_2, \cdots, r_n) = \sum_{i=2}^{n} l_i$$
(A.12)

$$= \sum_{i=2}^{n} \frac{r_i - r_{i-1} e^{-T_i/K}}{1 - e^{-T_i/K}} T_i$$

$$\leq \frac{r_2 - r_1 e^{-T_2/K}}{1 - e^{-T_2/K}} T_2 + \sum_{i=3}^{n} \frac{r_\alpha - r_\alpha e^{-T_i/K}}{1 - e^{-T_i/K}} T_i$$

$$= \frac{r_2 - r_1 e^{-T_2/K}}{1 - e^{-T_2/K}} T_2 + r_\alpha \sum_{i=3}^{n} T_i$$

$$= (r_2 - r_1) \frac{e^{-T_2/K}}{1 - e^{-T_2/K}} T_2 + r_2 T_2 + r_\alpha \sum_{i=3}^{n} T_i$$

$$< K r_\alpha + r_\alpha \sum_{i=2}^{n} T_i$$

$$\leq K r_\alpha + r_\alpha (t'' - t').$$

where the next to last inequality follows from the fact that $r_1 \geq 0$, $r_2 \leq r_\alpha$, and by using Eq. (A.2) after substitution $T_2 \to xK$.

By using Eq. (A.12), and the fact that $l_1 \leq l_{max}$, we get

$$\sum_{i=1}^{n} l_i = l_1 + F(r_1, r_2, \cdots, r_n) < l_{max} + K r_\alpha + r_\alpha (t'' - t'). \quad (A.13)$$

Since $(t'' - t') r_\alpha$ represents exactly the service to which the flow is entitled during the interval I, it follows that the excess service is bounded by $l_{max} + r_\alpha K$. $\square$

Lemma 2 *Consider a link with a normalized fair rate α, and a flow with weight w that sends at a rate no larger than R, where $R > r_\alpha$. Next consider an interval $I = [t', t'')$ such that t' represents the time just after a packet has arrived, and during which the rate estimator r is never smaller than $r_\alpha = w\alpha$, i.e., $r(t) \geq r_\alpha$, $\forall t \in I$. Then, the excess service received by the flow during I is bounded above by*

$$r_\alpha K \ln\frac{R}{r}. \quad (A.14)$$

Proof. Again, assume that during the interval I the flow sends exactly n packets. Similarly, let t_i be the arrival time of the i-th packet, and let l_i denote its length. Since we assume that when the packet i arrives, a *fraction* of that packet equal to flow i's forwarding probability, i.e., r_α/r_i is transmitted, the problem reduces to find an upper bound for

$$\sum_{i=1}^{n} l_i \frac{r_\alpha}{r_i}, \quad (A.15)$$

where $r_\alpha \leq r_i \leq R$, $1 \leq i \leq n$.

From Eq. (A.4), we obtain

$$\frac{l_i}{r_i} = \frac{r_i - r_{i-1} e^{-T_i/K}}{r_i(1 - e^{-T_i/K})} T_i \quad (A.16)$$

$$= T_i + \left(1 - \frac{r_{i-1}}{r_i}\right) \frac{e^{-T_i/K}}{1 - e^{-T_i/K}} T_i. \quad 1 \leq i \leq n.$$

Note that unlike Eq. (A.7), the above equation also applies for $i = 1$. This because we are guaranteed that there is at least one packet received before t' and therefore T_1 is well defined, i.e., from the hypothesis we have $T_1 = t_1 - t_0 = t_1 - t'$.

Further, by making substitution $x \to T_i/K$ in Eq. (A.2) we have

$$\frac{e^{-T_i/K}}{1 - e^{-T_i/K}} T_i < K, \quad T_i > 0. \tag{A.17}$$

From the above inequality and Eq. (A.16) we obtain

$$\frac{l_i}{r_i} < T_i + \left(1 - \frac{r_{i-1}}{r_i}\right) K, \quad 1 \le i \le n, \tag{A.18}$$

and further

$$\sum_{i=1}^{n} \frac{l_i}{r_i} = \sum_{i=1}^{n} T_i + K \sum_{i=1}^{n} \left(1 - \frac{r_{i-1}}{r_i}\right) \le (t'' - t') + K \sum_{i=1}^{n} \left(1 - \frac{r_{i-1}}{r_i}\right). \tag{A.19}$$

Next, since the arithmetic mean is no smaller than the geometric mean, i.e., $\left(\sum_{i=1}^{n} x_i\right)/n \ge \left(\prod_{i=1}^{n} x_i\right)^{1/n}$, where $x_i \ge 0$ $(1 \le i \le n)$, we have

$$\sum_{i=1}^{n} \left(1 - \frac{r_{i-1}}{r_i}\right) = n - \sum_{i=1}^{n} \frac{r_{i-1}}{r_i} \le n - n \left(\prod_{i=1}^{n} \frac{r_{i-1}}{r_i}\right)^{1/n} \tag{A.20}$$

$$= n \left(1 - \left(\frac{r_0}{r_n}\right)^{1/n}\right) \le n \left(1 - \left(\frac{r_\alpha}{R}\right)^{1/n}\right),$$

where the last inequality follows from the hypothesis, i.e., $r_\alpha \le r_i \le R$ $(0 \le i \le n)$.

By replacing $x \to (R/r_\alpha)^{1/n}$ in the well known inequality $\ln(x) > 1 - 1/x$, $(x > 1)$, we obtain $n(1 - (r_\alpha/R)^{1/n}) < \ln(R/r_\alpha)$, $n \ge 1$. Thus, Eq. (A.19) becomes

$$\sum_{i=1}^{n} \left(1 - \frac{r_{i-1}}{r_i}\right) < \ln \frac{R}{r_\alpha}. \tag{A.21}$$

Finally, from Eqs. (A.19) and (A.21) we obtain

$$\sum_{i=1}^{n} l_i \frac{r_\alpha}{r_i} \le r_\alpha(t'' - t') + r_\alpha K \sum_{i=1}^{n} \left(1 - \frac{r_{i-1}}{r_i}\right) \tag{A.22}$$

$$< r_\alpha(t'' - t') + r_\alpha K \ln \frac{R}{r_\alpha}$$

Since $(t'' - t')r_\alpha$ represents exactly the number of bits that the flow is entitled to send during the interval I, the proof follows. $\square$

Theorem 1 *Consider a link with a normalized fair rate α, and a flow with weight w. Then, the excess service received by a flow with weight w, that sends at a rate no larger than R, is bounded above by*

$$r_\alpha K \left(1 + ln\frac{R}{r_\alpha} \right) + l_{\max}, \tag{A.23}$$

where $r_\alpha = \alpha w$, and l_{max} represents the maximum length of the packet.

Proof. Assume the flow becomes active for the first time at t_a. Let t_b be the time when its rate estimator exceeds for the first time r_α, i.e., $r(t_b) > r_\alpha$ and $r(t) \leq r_\alpha$, $\forall t < t_b$. If such a time t_b does not exist, according to Lemma 1, the excess service received by the flow is bounded by $r_\alpha K + l_{\max}$, which concludes the proof for this case. In the following paragraphs, we consider the case when t_b exists.

Next, we show that the service received by the flow is maximized when $r(t) \geq r_\alpha$, $\forall t > t_b$. The proof is by contradiction. Assume there is an interval $I' = [t', t'') \subset I$, such that $t' \geq t_b$, and that $r(t) < r_\alpha$, $(t' \leq t < t'')$. Then using an identical argument as in Lemma 1, it can be shown that the service that the flow receives during I' increases when $r(t) = r_\alpha$, $\forall t \in I'$. The only change in the proof of Lemma 1 is that now Eq. (A.7) will also apply for $i = 1$, as according to the hypothesis the estimated rate just before t' (i.e., r_0 in Lemma 1) is greater than zero; more precisely $r_0 \geq r_\alpha$. Further, by including l_1 in the definition of $F(\cdot)$ (see Eq. (A.8)) we show that $F(\cdot)$ is increasing in each of its arguments r_i, $1 \leq i \leq n$.

Thus, the service received by the flow is maximized when the estimated rate of the flow is no smaller than r_α after time t_b. But then, according to Lemma 2, the excess service received by the flow after t_b is bounded by [1]

$$r_\alpha K ln\frac{R}{r_\alpha}. \tag{A.24}$$

Similarly, from Lemma 1 it follows that the excess service received by the flow during the interval $[t_a, t_b)$ is bounded above by

$$r_\alpha K + l_{max}, \tag{A.25}$$

and therefore by combining (A.24) and (A.25) the total excess service is bounded above by

$$r_\alpha K \left(1 + ln\frac{R}{r_\alpha} \right) + l_{max}. \tag{A.26}$$

$\square$

---

[1]Without loss of generality here we assume that t_b represents the time just after r was evaluated as being smaller than r_α for the last time. Since this coincides with a packet arrival Lemma 2 applies.

B Performance Bounds for Guaranteed Services

B.1 Network Utilization of Premium Service in Diffserv Networks

Premium service provides the equivalent of a dedicated link of fixed bandwidth between edge nodes in a Diffserv network. In such a service, each premium flow has a reserved peak rate. In the data plane, ingress nodes police each premium service traffic flow according to its peak reservation rate. Inside the Diffserv domain, core routers put the aggregate of all premium traffic into one scheduling queue and service the premium traffic with strict priority over best effort traffic. In the control plane, a bandwidth broker is used to perform admission control. The idea is that by using very conservative admission control algorithms based on worst case analysis, together with peak rate policing at ingress nodes and static priority scheduling at core nodes, it is possible to ensure that all premium service packets incur very small queueing delay.

One important design question to ask is how conservative does the admission control algorithm need to be? In other words, what is the upper limit on the utilization of the network capacity that can be allocated to premium traffic if we want the premium service to achieve the same level of service assurance as the guaranteed service, such that the queueing delay of all premium service packets is bounded by a fixed number even in the worst case?

For the purpose of this discussion, we use *flow* to refer to a subset of packets that traverse the same path inside a Diffserv domain between two edge nodes. Thus, with the highest level of traffic aggregation, a flow consists of all packets between the same pair of ingress and egress nodes. Note that even in this case, the number of flows in a network can be quite large as this number may increase quadratically with the number of edge nodes.

Let us consider a domain consisting of 4×4 routers with links of capacity C. Assume that the fraction of the link capacity allocated to the premium traffic is limited to γ. Assume also that all flows have equal packet sizes, and that each ingress node shapes not only each flow, but also the aggregate traffic at each of its outputs. Figure B.1(a) shows the traffic pattern at the first core router along a path. Each input receives 12 identical flows, where each flow has a reservation of $\gamma C/12 = C/48$. Let τ be the transmission time

I. Stoica: Stateless Core, LNCS 2979, pp. 191-212, 2004.
© Springer-Verlag Berlin Heidelberg 2004

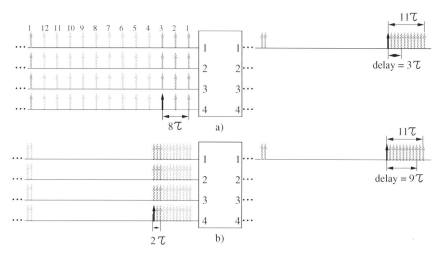

Fig. B.1. Per-hop worst-case delay experienced by premium traffic in a Diffserv domain. (a) and (b) show the traffic pattern at the first and a subsequent node. The black and all dark grey packets go to the first output; the light grey packets go to the other outputs.

of one packet, then as shown in the Figure, the inter-arrival time between two consecutive packets in the each flow is 48τ, and the inter-arrival time between two consecutive packets in the aggregate flow is 4τ.

Assume the first three flows at each input are forwarded to output 1. This will cause a burst of 12 packets to arrive at output 1 in a 8τ long interval and the last packet of the burst to incur an *additional* delay of 3τ. Now assume that the next router receives at each input a traffic pattern similar to the one generated by output 1 of the first core router, as shown in Figure B.1(b). In addition, assume that the *last* three flows from each input burst are forwarded to output 1. This will cause a burst of 12 packets to arrive 1 at output in a 2τ long interval and the last packet in the burst to incur an *additional* delay of 9τ. Thus, after two hops, a packet is delayed by as much as 12τ. This pattern can be repeated for all subsequent hops.

In general, consider a $k \times k$ router, and let n be the number of flows that traverse each link. For simplicity, assume that $\gamma \geq 1/k$. Then it can be shown that the worst case delay experienced by a packet after h hops is

$$D = \left(n - 1 - \left(\frac{n}{k} - 1\right)\frac{1}{\gamma}\right)\tau + (h-1)n\frac{k-1}{k}\tau + h\tau, \qquad \text{(B.1)}$$

where the first term is the additional delay at the first hop, the second term is the additional delay at all subsequent hops, and the last term accounts for the packet transmission time at each hop. As a a numerical example, let $C = 1$ Gbps, a packet size of 1500 bytes, $k = 16$, $\gamma = 10\%$, $n = 1500$ and $h = 15$. From here we obtain $\tau = 12$ μsec, and a delay D of over 240 ms.

Finally, if $\gamma < 1/k$, it can be shown that it will take only $\lceil \log_k(1/\gamma) \rceil$ hops to achieve a continuous burst. For example, for $\gamma = 1\%$ and $k = 16$, it takes only two hops to obtain a continuous burst.

The above example demonstrates that low network utilization and traffic shaping at ingress nodes alone are not enough to guarantee a "small" worst-case delay for *all* the premium traffic. This result is not surprising. Even using a per flow scheduler like Weighted Fair Queueing (WFQ), will not help to reduce the worst case end-to-end delay for *all* packets. In fact, if all flows in the above example are given the same weight, the worst case delay under WFQ is $hn\tau$, which is basically the same as the one given by Eq. (B.1). However, the major advantage of using WFQ is that it allows us to *differentiate* among flows, which is a critical property as long as we cannot guarantee a "small" delay to all flows. In addition, WFQ can achieve 100% utilization.

B.2 Proof of Theorem 2

In this appendix we show that a network of CJVC servers provides the same end-to-end delay guarantees as a network of Jitter-VC servers. In particular, in Theorem 2 we show that the deadline of a packet at the last hop in both systems is the same. This result is based on Lemmas 4 and 5 which give the expressions of the deadline of a packet at the last hop in a network of Jitter-VC, and a network of CJVC servers, respectively. First, we present a preliminary result used in proving Lemma 4.

Lemma 3 *Consider a network of Jitter-VC servers. Let π_j denote the propagation delay between hops j and $j+1$, and let τ_j be the maximum transmission time of a packet at node j. Then for any $j > 1$ and $i, k \geq 1$ we have*

$$d_{i,j+1}^k - d_{i,j}^k - \tau_j - \pi_j \geq d_{i,j}^k - d_{i,j-1}^k - \tau_{j-1} - \pi_{j-1}. \qquad (B.2)$$

Proof. The proof is by induction on k. First, recall that by definition $g_{i,j}^k = d_{i,j}^k + \tau_j - s_{i,j}^k$ (see Table 5.1), and that for $j > 1$, $a_{i,j}^k = s_{i,j-1}^k + \pi_{j-1}$. From here and from Eqs. (5.1) and (5.2) we have then

$$d_{i,j}^k = \max(a_{i,j}^k + g_{i,j-1}^k, d_{i,j}^{k-1}) + \frac{l_i^k}{r_i} \qquad (B.3)$$

$$= \max(d_{i,j-1}^k + \tau_{j-1} + \pi_{j-1}, d_{i,j}^{k-1}) + \frac{l_i^k}{r_i}.$$

Basic Step. For $k = 1$ and any $j \geq 1$, from Eq. (B.3) we have trivially $d_{i,j}^1 = d_{i,j-1}^1 + \tau_{j-1} + \pi_{j-1} + l_i^1/r_i, \forall j > 1$, and therefore $d_{i,j}^1 - d_{i,j-1}^1 - \tau_{j-1} - \pi_{j-1} = l_i^1/r_i, \forall j > 1$.

Induction Step. Assume Eq. (B.2) is true for k. Then we need to show that

$$d_{i,j+1}^{k+1} - d_{i,j}^{k+1} - \tau_j - \pi_j \geq d_{i,j}^{k+1} - d_{i,j-1}^{k+1} - \tau_{j-1} - \pi_{j-1} \Rightarrow \qquad (B.4)$$
$$\max(d_{i,j}^{k+1} + \tau_j + \pi_j, d_{i,j+1}^k) - \max(d_{i,j-1}^{k+1} + \tau_{j-1} + \pi_{j-1}, d_{i,j}^k) - \tau_j - \pi_j \geq$$
$$\max(d_{i,j-1}^{k+1} + \tau_{j-1} + \pi_{j-1}, d_{i,j}^k) - \max(d_{i,j-2}^{k+1} + \tau_{j-2} + \pi_{j-2}, d_{i,j-1}^k) - \tau_{j-1} - \pi_{j-1},$$

where the second inequality follows after using Eq. (B.3). Next consider two cases: whether $d_{i,j-1}^{k+1} + \tau_{j-1} + \pi_{j-1} \leq d_{i,j}^k$ or not. Assume $d_{i,j-1}^{k+1} + \tau_{j-1} + \pi_{j-1} \leq d_{i,j}^k$. From Eq. (B.4) and from the induction hypothesis we obtain

$$
\begin{aligned}
d_{i,j+1}^{k+1} - d_{i,j}^{k+1} - \tau_j - \pi_j &= \max(d_{i,j}^{k+1} + \tau_j + \pi_j, d_{i,j+1}^k) - && (B.5) \\
&\quad \max(d_{i,j-1}^{k+1} + \tau_{j-1} + \pi_{j-1}, d_{i,j}^k) - \tau_j - \pi_j \\
&= \max(d_{i,j}^{k+1} + \tau_j + \pi_j, d_{i,j+1}^k) - d_{i,j}^k - \tau_j - \pi_j \\
&\geq d_{i,j+1}^k - d_{i,j}^k - \tau_j - \pi_j && \text{(induction hypothesis)} \\
&\geq d_{i,j}^k - d_{i,j-1}^k - \tau_{j-1} - \pi_{j-1} \\
&\geq d_{i,j}^k - \max(d_{i,j-2}^{k+1} + \tau_{j-1} + \pi_{j-1}, d_{i,j-1}^k) - \tau_{j-1} - \pi_{j-1} \\
&= \max(d_{i,j-1}^{k+1} + \tau_{j-1} + \pi_{j-1}, d_{i,j}^k) - \\
&\quad \max(d_{i,j-2}^{k+1} + \tau_{j-2} + \pi_{j-2}, d_{i,j-1}^k) - \tau_{j-1} - \pi_{j-1} \\
&= d_{i,j}^{k+1} - d_{i,j-1}^{k+1} - \tau_{j-1} - \pi_{j-1}.
\end{aligned}
$$

Next, assume that

$$d_{i,j-1}^{k+1} + \tau_{j-1} + \pi_{j-1} > d_{i,j}^k. \qquad (B.6)$$

From here and by using Eq. (B.3) and Eq. (B.4) we have

$$
\begin{aligned}
d_{i,j+1}^{k+1} - d_{i,j}^{k+1} - \tau_j - \pi_j &= \max(d_{i,j}^{k+1} + \tau_j + \pi_j, d_{i,j+1}^k) - && (B.7) \\
&\quad \max(d_{i,j-1}^{k+1} + \tau_{j-1} + \pi_{j-1}, d_{i,j}^k) - \tau_j - \pi_j \\
&= \max(d_{i,j}^{k+1} + \tau_j + \pi_j, d_{i,j+1}^k) - d_{i,j-1}^{k+1} - \\
&\quad \tau_{j-1} - \pi_{j-1} - \tau_j - \pi_j && (B.8) \\
&\geq d_{i,j}^{k+1} - d_{i,j-1}^{k+1} - \tau_{j-1} - \pi_{j-1} \\
&= d_{i,j}^{k+1} - \max(d_{i,j-1}^{k+1} + \tau_{j-1} + \pi_{j-1}, d_{i,j}^k) \\
&= \frac{l_i^{k+1}}{r_i} && \text{(from Eq. (B.3))} \\
&= d_{i,j-1}^{k+1} - \max(d_{i,j-2}^{k+1} + \tau_{j-2} + \pi_{j-2}, d_{i,j-1}^k) \\
&= \max(d_{i,j-1}^{k+1} + \tau_{j-1} + \pi_{j-1}, d_{i,j}^k) - \tau_{j-1} - \pi_{j-1} -
\end{aligned}
$$

$$\max(d_{i,j-2}^{k+1} + \tau_{j-2} + \pi_{j-2}, d_{i,j-1}^k) \quad \text{(from Eq. (B.6))}$$
$$= d_{i,j}^{k+1} - d_{i,j-1}^{k+1} - \tau_{j-1} - \pi_{j-1}.$$

This completes the proof. $\square$

Lemma 4 *The deadline of any packet* p_i^k, $k > 1$, *at the last hop* h *in a network of Jitter-VC servers is*

$$d_{i,h}^k = \max\left(e_{i,1}^k + h\frac{l_i^k}{r_i} + \sum_{m=1}^{h-1}(\tau_m + \pi_m), d_{i,h}^{k-1} + \frac{l_i^k}{r_i} \right). \tag{B.9}$$

Proof. Let $j^* > 1$ be the last hop for which $d_{i,j^*-1}^k + \tau_{j^*-1} + \pi_{j^*-1} < d_{i,j^*}^{k-1}$. We consider two cases whether j^* exists or not.

Case 1. (j^* does not exist) From Eq. (5.1) we have $e_{i,j}^k = d_{i,j-1}^k + \tau_{j-1} + \pi_{j-1}$, $\forall j > 1$. From here and by using Eq. (5.2) we obtain

$$d_{i,h}^k = e_{i,1}^k + h\frac{l_i^k}{r_i} + \sum_{m=1}^{h-1}(\tau_m + \pi_m). \tag{B.10}$$

Because we assume that j^* does not exist we also have $d_{i,h}^k = e_{i,h}^k + l_i^k/r_i \geq d_{i,h}^{k-1} + l_i^k/r_i$, which concludes the proof of this case.

Case 2. (j^* exists) In this case we show that $j^* = h$. Assume this is not true. Then we have $e_{i,j}^k = d_{i,j-1}^k + \tau_{j-1} + \pi_{j-1}$, $\forall j > j^*$. By using Eq. (5.2) we obtain

$$d_{i,h}^k = e_{i,j^*}^k + (h - j^* + 1)\frac{l_i^k}{r_i} + \sum_{m=j^*}^{h-1}(\tau_m + \pi_m). \tag{B.11}$$

On the other hand, by the definition of j^* and from Eqs. (5.1) and (5.2) we have

$$d_{i,j^*}^k = \max(d_{i,j^*-1}^k + \tau_{j^*-1} + \pi_{j^*-1}, d_{i,j^*}^{k-1}) + \frac{l_i^k}{r_i} \tag{B.12}$$

$$> d_{i,j^*-1}^k + \tau_{j^*-1} + \pi_{j^*-1} + \frac{l_i^k}{r_i}.$$

As a result we obtain $d_{i,j^*}^k - d_{i,j^*-1}^k - \tau_{j^*-1} - \pi_{j^*-1} > l_i/r_i$. By iteratively applying Lemma 3 we have

$$d_{i,m+1}^k - d_{i,m}^k - \tau_m - \pi_m \geq d_{i,j^*}^k - d_{i,j^*-1}^k - \tau_{j^*-1} - \pi_{j^*-1} > \frac{l_i^k}{r_i}, \text{(B.13)}$$

$\forall m \geq j^*$. From Eq. (B.13) we obtain

$$\sum_{m=j^*}^{h-1} (d_{i,m+1}^k - d_{i,m}^k - \tau_m - \pi_m) \geq (h - j^*)(d_{i,j^*}^k - d_{i,j^*-1}^k - \tau_{j^*-1} - \pi_{j^*-1})$$

$$> (h - j^*)\frac{l_i^k}{r_i}, \tag{B.14}$$

where the right-hand term can be expressed as

$$\sum_{m=j^*}^{h-1} (d_{i,m+1}^k - d_{i,m}^k - \tau_m - \pi_m) = d_{i,h}^k - d_{i,j^*}^k - \sum_{m=j^*}^{h-1} (\tau_m + \pi_m). \text{(B.15)}$$

By combining Eq. (B.14) and Eq. (B.15) we get

$$d_{i,h}^k > d_{i,j^*}^k + (h - j^*)\frac{l_i^k}{r_i} + \sum_{m=j^*}^{h-1} (\tau_m + \pi_m) \tag{B.16}$$

$$= e_{i,j^*}^k + (h - j^* + 1)\frac{l_i^k}{r_i} + \sum_{m=j^*}^{h-1} (\tau_m + \pi_m).$$

But this inequality contradicts Eq. (B.11) and therefore proves our statement, i.e., $j^* = h$. Thus, $e_{i,h}^k = d_{i,h}^{k-1}$. From here and from Eqs. (5.1) and (5.2) we get

$$d_{i,h}^k = e_{i,h}^k + \frac{l_i}{r_i} = d_{i,h}^{k-1} + \frac{l_i}{r_i}. \tag{B.17}$$

Now, from Eq. (5.1) it follows trivially that

$$e_{i,j}^k = \max(d_{i,j-1}^k + \tau_{j-1} + \pi_{j-1}, d_{i,j-1}^k) \tag{B.18}$$

$$\geq d_{i,j-1}^k + \tau_{j-1} + \pi_{j-1}, \quad j \geq 1.$$

By iterating over the above equation and then using Eq. (5.2) we get

$$d_{i,h}^k \geq e_{i,1}^k + h\frac{l_i^k}{r_i} + \sum_{m=1}^{h-1} (\tau_m + \pi_m), \tag{B.19}$$

which, together with Eq. (B.17), lead us to Eq. (B.9).

This completes the proof of the lemma. □

Lemma 5 *The deadline of any packet p_i^k, $k > 1$, at the last hop h in a network of CJVC servers is*

$$d_{i,h}^k = \max\left(e_{i,1}^k + h\frac{l_i^k}{r_i} + \sum_{m=1}^{h-1} (\tau_m + \pi_m), d_{i,h}^{k-1} + \frac{l_i^k}{r_i}\right). \tag{B.20}$$

Proof. We consider two cases whether $\delta_i^k = 0$ or not.

Case 1. ($\delta_i^k = 0$) From Eqs. (5.2) and (5.6) it follows that

$$d_{i,h}^k = e_{i,1}^k + h\frac{l_i^k}{r_i} + \sum_{m=1}^{h-1}(\tau_m + \pi_m). \tag{B.21}$$

On the other hand, by the definition of δ_i^k (see Eq. (5.3) and Eq. (5.4)) we have $e_{i,j}^k = d_{i,j-1}^k + \tau_{j-1} + \pi_{j-1} + \delta_i^k \geq d_{i,j}^{k-1}$, $\forall j > 1$. From here and from Eq. (5.2) we obtain

$$d_{i,h}^k \geq d_{i,h}^{k-1} + \frac{l_i^k}{r_i}. \tag{B.22}$$

From this inequality and Eq. (B.21), Eq. (B.20) follows.

Case 2. ($\delta_i^k > 0$) By using Eqs. (5.2) and (5.10) we obtain

$$d_{i,h}^k = e_{i,1}^k + h\frac{l_i^k}{r_i} + (h-1)\delta_i^k + \sum_{m=1}^{h-1}(\tau_m + \pi_m) \tag{B.23}$$

$$= e_{i,1}^k + h\frac{l_i^k}{r_i} + \left((h-1)\delta_i^{k-1} + (h-1)\frac{l_i^{k-1} - l_i^k}{r_i} - e_{i,1}^k + e_{i,1}^{k-1} + \frac{l_i^{k-1}}{r_i}\right)$$

$$\sum_{m=1}^{h-1}(\tau_m + \pi_m)$$

$$= e_{i,1}^{k-1} + h\frac{l_i^{k-1}}{r_i} + (h-1)\delta_i^{k-1} + \sum_{m=1}^{h-1}(\tau_m + \pi_m) + \frac{l_i^k}{r_i}$$

$$= d_{i,h}^{k-1} + \frac{l_i^k}{r_i}.$$

Since $\delta_i^k > 0$, by using again Eq. (5.2) and (5.7) we get

$$d_{i,h}^k = e_{i,1}^k + h\frac{l_i^k}{r_i} + (h-1)\delta_i^k + \sum_{m=1}^{h-1}(\tau_m + \pi_m) \tag{B.24}$$

$$> e_{i,1}^k + h\frac{l_i^k}{r_i} + \sum_{m=1}^{h-1}(\tau_m + \pi_m).$$

which, together with Eq. (B.23), lead to Eq. (B.20). $\square$

Theorem 2 *The deadlines of a packet at the last hop in a network of CJVC servers is equal to the deadline of the same packet in a corresponding network of Jitter-VC servers.*

Proof. From Eqs (5.1) and (5.2) it is easy to see that in a network of Jitter-VC servers we have

$$d_{i,h}^1 = e_{i,1}^1 + h \frac{l_i^k}{r_i} + \sum_{m=1}^{h-1} (\tau_m + \pi_m). \tag{B.25}$$

Similarly, in a network of CJVC servers, from Eqs. (5.1) and (5.7), and by using the fact that $\delta_i^1 = 0$ (see Eq. 5.8), we obtain an identical expression for $d_{i,h}^1$ (i.e., Eq. (B.25)).

Finally, since (a) the eligible times of all packets p_i^k at the first hop, i.e., $e_{i,1}^k$ ($\forall k \geq 1$), are identical for both Jitter-VC and CJVC servers, and since (b) the deadlines of the packets at the last hop, i.e., $d_{i,h}^k$ ($\forall k \geq 1$), are computed based on the same formulae (see Eqs. (B.9), (B.20) and B.25), it follows that $d_{i,h}^k$, ($\forall k \geq 1$) are identical in both a network of Jitter-VC, and a network of CJVC servers. $\square$

B.3 Proof of Theorem 3

To prove Theorem 3 (see Section 5.3.3) we prove two intermediate results: Lemma 9 which gives the buffer occupancy for the case when all flows have identical rates, and Lemma 12 which gives the buffer occupancy for arbitrary flow rates.

B.3.1 Identical Flow Rates

Consider a work-conserving server with an output rate of one, which is traversed by n flows with identical reservations of $1/n$. Assume that the time axis is divided in *unit* sized slots, where slot t corresponds to the time interval $[t, t+1)$. Assume that at most one packet can be sent during each slot, i.e., the packet transmission time is one time unit. Finally, assume that the starting times of the backlogged periods of any two flows are uncorrelated. In practice, we enforce this by delaying the first packet of a backlogged period by an amount drawn from a uniform distribution in the range $[t_{arrival}, t_{arrival} + n)$, where $t_{arrival}$ is the arrival time of the first packet in the backlogged period. Note that according to Eq. (5.1), the eligible times of the packets of a flow during a flow's backlogged interval are *periodic* with period n. Thus, without loss of generality, we assume that the arrival process of any flow during a backlogged interval is periodic.

Let $r(t', t'')$ denote the number of packets received (i.e., became eligible) during the interval $[t', t'')$, and let $s(t', t'')$ denote the number of packets sent during the same interval. Note that $r(t', t'')$ and $s(t', t'')$ do *not* include packets received/transmitted during slot t''. Let $q(t)$ denote the size of the queue at the *beginning* of slot t. Then, if no packets are dropped, we have

$$q(t'') = q(t') + r(t', t'') - s(t', t''). \tag{B.26}$$

Since at most one packet is sent during each time slot, we have $s(t', t'') \leq t'' - t'$. The inequality holds when $[t', t'')$ belongs to a server busy period. A busy period is defined as an interval during which the server's queue is never empty. Also, note that if t' is the starting time of a busy period $q(t') = 0$.

The next result shows that to compute an upper bound for $q(t)$, it is enough to consider only the scenarios in which all flows are continuously backlogged.

Lemma 6 *Let t_1 be an arbitrary time slot during a server busy period that starts at time t_0. Assume flow i is not continuously backlogged during the interval $[t_0, t_1)$. Then $q(t_1)$ can only increase if flow i becomes continuously backlogged during $[t_0, t_1)$.*

Proof. Consider two cases in which flow i is idle during the entire interval $[t_0, t_1)$, and not.

If flow i is idle during $[t_0, t_1)$, consider the modified scenario in which flow i becomes backlogged at an arbitrary time $t < t_0$, and remains continuously backlogged during $[t_0, t_1)$. In addition, assume that the arrival patterns of all the other flows remain unchanged. As a result, it is easy to see that in the modified scenario, the total number of packets received during $[t_0, t_1)$ can only increase, while the starting time of the busy interval can only decrease. Let r', s', and q' denote the corresponding values in the modified scenario. Then $q'(t_0) \geq q(t_0) = 0$, $r'(t_0, t_1) \geq r(t_0, t_1)$, and $s'(t_0, t_1) = s(t_0, t_1) = t_1 - t_0$. From Eq. (B.26) it follows then that $q'(t_1) \geq q(t_1)$.

In the second case, when flow i is neither idle nor continuously backlogged during the interval $[t_0, t_1)$, let t' denote the time when the last packet of flow i arrives during $[t_0, t_1)$. Next consider the modified scenario in which flow i's packets arrive at times: $t' - na, \ldots, t' - n, t', t' + n, \ldots, t' + nb$, such that $t' - na \leq t_0$, and $t_1 \leq t' + nb$. It is easy to see then that the number of packets of flow i that arrive during $[t_0, t_1)$ is no smaller than the number of packets of flow i that arrive during the same interval in the original scenario. By assuming that the arrival patterns of all the other flows do not change, it follows that $r'(t_0, t_1) \geq r(t_0, t_1)$. In addition, since at most $t_1 - t_0$ packets are transmitted during $[t_0, t_1)$ we have $s'(t_0, t_1) \leq t_1 - t_0$. The inequality holds if, after changing the arrival pattern of flow i, the server is no longer busy during the entire interval $[t_0, t_1)$. In addition, we have $q'(t_0) \geq 0$, and from the hypothesis $q(t_0) = 0$. Finally, from Eq. (B.26) we obtain $q'(t_1) \geq q(t_1)$, which concludes the proof of the lemma. $\square$

As a consequence, in the remainder of this section, we limit our study to a busy period in which all flows are continuously backlogged.

Let t_1 be the time when the last flow becomes backlogged. Let t_0 be the latest time no larger than t_1 when the server become busy, i.e., it has no packet to send during $[t_0 - 1, t_0)$ and is continuously busy during the interval $[t_0, t_1 + 1)$. Then we have the following result.

Lemma 7 *If all flows remain continuously backlogged after time t_1, the server is busy for any time $t \geq t_0$.*

Proof. By the definition of t_0, the server is busy during $[t_0, t_1)$. Next we show that the server is also busy for any $t_1 \geq 0$.

Consider a flow that becomes backlogged at time t', Since its arrival process is periodic it follows that during any interval $[t' - n + i, t' + i)$, $\forall i > 0$, exactly one packet of this flow arrives. Since after time t_1 all n flows are backlogged, exactly n packets are received during $[t_1 - n + i, t_1 + i)$, $\forall i > 0$. Since at most n packets are sent during each of these intervals, it follows that the server cannot be idle during any slot i. $\square$

Consider a buffer of size s. Our goal is to compute the probability that the buffer will overflow during an arbitrary interval $[t_0, t_0 + d)$. From Lemma 7 it follows that since the server is busy during $[t_0, t_0 + d)$, exactly d packets are transmitted during this interval. In addition, since the starting times of the backlogged periods of different flows are not correlated, in the remainder of this section we also assume that the starting times of a flow's backlogged period is *not* correlated with the starting time, t_0, of a busy period. Thus, during the interval $[t_0, t_0 + d)$, a flow receives $\lceil d/n \rceil$ packets with probability $p(d) = d/n - \lfloor d/n \rfloor$, and $\lfloor d/n \rfloor$ with probability $1 - p$. Since this probability is periodic with period n it will suffice to consider only intervals of size equal to at most n. Consequently, we will assume $d \leq n$. The probability to receive one packet during $[t_0, t_0 + d)$ is then

$$p(d) = \frac{d}{n}. \tag{B.27}$$

Let $p(m; d)$ denote the probability with which exactly m packets are received during the time interval $[t_0, t_0 + d)$, where

$$p(m; d) = \binom{n}{m} p(d)^m (1 - p(d))^{n-m}. \tag{B.28}$$

Now, let $P(x > s, u)$ denote the probability with which the queue size exceeds s *at* time $t_0 + u$. Since the server is idle at t_0 and busy during $[t_0, t_0 + u)$, from Eq. (B.26), it follows that the server's queue overflows when more than $u + s$ packets are received during $[t_0, t_0 + u)$. Thus, we have

$$P(x > s, u) = \sum_{i=u+s+1}^{n} p(i; u) = \sum_{i=u+s+1}^{n} \binom{n}{i} p(u)^i (1 - p(u))^{n-i}. \tag{B.29}$$

The next result computes $P(x > s, u)$.

Lemma 8 *The probability that a queue of size s overflows at time $t_0 + u$ is bounded by*

$$P(x > s, u) < \beta(n) \sqrt{\frac{1}{2\pi} \left(\frac{1 - (s-1)/2n}{1 + (s-1)/2n} \right)^{2s} \frac{(n+s)^2}{4sn}}. \tag{B.30}$$

where $\beta(n) = (n/e)^{1+(1/12n)}$.

Proof. From Eq. (B.28) we obtain

$$p(m+1; u) = \frac{p(u)}{1 - p(u)} \cdot \frac{n - m}{m + 1} \cdot p(m; u), \tag{B.31}$$

By plugging the above equation and Eq. (B.27) into Eq. (B.29) we obtain

$$P(x > s, u) = p(u + s; u) \sum_{i=u+s+1}^{n} \left(\prod_{k=u+s}^{i-1} \frac{n - k}{k + 1} \right) \left(\frac{u}{n - u} \right)^{i-u-s} \tag{B.32}$$

$$< p(u + s; u) \sum_{i=u+s+1}^{n} \left(\prod_{k=u+s}^{i-1} \frac{n - u - s}{u + s} \right) \left(\frac{u}{n - u} \right)^{i-u-s}$$

$$= p(u + s; u) \sum_{i=u+s+1}^{n} \left(\frac{n - u - s}{u + s} \cdot \frac{u}{n - u} \right)^{i-u-s}$$

Next, it can be easily verified that for any positive reals a, b, and x, such that $b - x \geq 0$, we have

$$\frac{a}{a + x} \cdot \frac{b - x}{b} \leq \left(\frac{a + b - x}{a + b + x} \right)^2. \tag{B.33}$$

By taking $a = u$, $b = n - u$, $x = s$, Eq. (B.32) becomes

$$P(x > s, u) < p(u + s; u) \sum_{i=u+s+1}^{n} \left(\frac{n - s}{n + s} \right)^{2(i-u-s)} \tag{B.34}$$

$$< p(u + s; u) \sum_{i=0}^{\infty} \left(\frac{n - s}{n + s} \right)^{2i}$$

$$< p(u + s; u) \frac{(n + s)^2}{4sn}.$$

Next, it remains to bound $p(u + s; u)$. From Eqs. (B.27) and (B.31) we have

$$p(u + s; u) = p(u; u) \prod_{i=0}^{s-1} \left(\frac{u}{n - u} \cdot \frac{n - u - i}{u + i + 1} \right) \tag{B.35}$$

$$< \prod_{i=0}^{s-1} \left(\frac{u}{u + i} \cdot \frac{n - u - i}{n - u} \right).$$

By using Eq. (B.33) with $a = u$, $b = n - u$, and $x = i$, we obtain

$$p(u + s; u) = p(u; u) \prod_{i=0}^{s-1} \left(\frac{n - i}{n + i} \right)^2 \tag{B.36}$$

$$= p(u; u) \prod_{i=0}^{s-1} \left(\frac{n - i}{n + i} \cdot \frac{n - s + 1 + i}{n + s - 1 - i} \right).$$

Again, by applying Eq. (B.33) to the pairs $(n-i)/(n+i)$ and $(n-s+1+i)/(n+s-1-i)$, $\forall i < s/2$, we have

$$p(u+s;u) < p(u;u)\left(\frac{2n-(s-1)}{2n+(s-1)}\right)^{2s} = p(u;u)\left(\frac{1-(s-1)/2n}{1+(s-1)/2n}\right)^{2s} \quad \text{(B.37)}$$

To bound $p(u;u)$ we use Stirling inequalities [26], i.e., $\sqrt{2\pi n}(n/e)^n < n! < \sqrt{2\pi n}(n/e)^{n+(1/12n)}$, $\forall n \geq 1$. From here we have

$$\binom{n}{n-u} < \frac{\sqrt{2\pi n}(n/e)^{n+(1/12n)}}{\sqrt{2\pi u}(u/e)^u\sqrt{2\pi(n-u)}((n-u)/e)^{n-u}} \quad \text{(B.38)}$$

$$= \sqrt{\frac{n}{2\pi(n-u)u}} \cdot \frac{n^n(n/e)^{1/12n}}{u^u(n-u)^{n-u}}.$$

By combining Eqs. (B.27), (B.28) and (B.38), we obtain

$$p(u;u) < \beta(n)\sqrt{\frac{n}{2\pi u(n-u)}} \leq \beta(n)\sqrt{\frac{1}{2\pi}}. \quad \text{(B.39)}$$

where $\beta(n) = (n/e)^{1+(1/12n)}$ and the last inequality follows from the fact that $n/((n-u)u) < 1$, for any $u \geq 1$, $n \geq 2$. By plugging the above result in Eq. (B.34) we obtain

$$P(x > s, u) < \beta(n)\sqrt{\frac{1}{2\pi}}\left(\frac{1-(s-1)/2n}{1+(s-1)/2n}\right)^{2s}\frac{(n+s)^2}{4sn}. \quad \text{(B.40)}$$

□

Lemma 9 *Consider n flows with identical rates and unit packet sizes. Then given a buffer of size s, where*

$$s \geq \sqrt{n\left(\frac{\ln n}{2} - \frac{\ln \varepsilon}{2} - 1\right)}, \quad \text{(B.41)}$$

the probability that the buffer overflows during an arbitrary time slot when the server is busy is asymptotically $< \varepsilon$.

Proof. To compute the asymptotic bound for $P(x > s, u)$ assume that $s \ll n$. Since $(1-x)/(1+x) \simeq 1 - 2x$ and $\ln(1-x) \simeq x$, for $x \to 0$, and since $(n+s)^2/sn < n$ for $n > s \geq 4$, and $\beta(n) < 1.102$ for any $n \geq 1$, by using Eq. (B.30) we obtain[1]

---

[1]More precisely $\ln \beta(n)\sqrt{1/(2\pi)} - \ln 4 \leq -2.2081062\ldots$.

$$\ln P(x > s, u) \simeq \ln \left(\beta(n) \sqrt{\frac{1}{2\pi}} \right) + 2s \cdot \ln \left(\frac{1 - (s-1)/2n}{1 + (s-1)/2n} \right) + \qquad \text{(B.42)}$$

$$\ln n - \ln 4$$

$$\simeq \ln \left(\beta(n) \sqrt{\frac{1}{2\pi}} \right) + 2s \cdot \ln \left(1 - \frac{s-1}{n} \right) + \ln n - \ln 4$$

$$\simeq \ln \left(\beta(n) \sqrt{\frac{1}{2\pi}} \right) - 2s \frac{s-1}{n} + \ln n - \ln 4$$

$$< -2 - 2 \frac{s(s-1)}{n} + \ln n \simeq 2(-1 - \frac{s^2}{n}) + \ln n.$$

Using ε to bound $P(x > s, u)$ leads us to

$$P(x > s; u) \leq \varepsilon \Rightarrow \qquad \qquad \text{(B.43)}$$

$$2 \left(-1 - \frac{s^2}{n} \right) + \ln n \leq \ln \varepsilon \Rightarrow$$

$$s \geq \sqrt{n \left(\frac{\ln n}{2} - \frac{\ln \varepsilon}{2} - 1 \right)}.$$

□

Next we prove a stronger result by computing an asymptotic upper bound for the probability with which a queue of size s overflows during an *arbitrary* busy interval. Let $Q(x > s)$ denote this probability. The key observation is that since all flows have period n, the aggregate arrival traffic will have the same period n. In addition, since during each of these periods exactly n packets are received/transmitted it follows that the queue size at any time $t_0 + i \cdot n + j$ is the same, $\forall i, j \geq 0$. Consequently, if the queue does not overflow during $[t_0, t_0 + n)$, the queue will not overflow at any other time $t \geq t_1$ during the same busy period. Thus, the problem reduces to compute the probability of queue overflowing during the interval $[t_0, t_0 + n)$. Then we have the following result.

Corollary 1 *Consider n flows with identical rates and unit packet sizes. Then given a buffer of size s, where*

$$s \geq \sqrt{n(\ln n - (\ln \varepsilon)/2 - 1)}, \qquad \qquad \text{(B.44)}$$

the probability that the buffer overflows during an arbitrary busy interval is asymptotically $< \varepsilon$.

Proof. Let ε' be the probability that a buffer of size s overflows at an instant t during the busy interval $[t_0, t_0 + u)$. Then the probability that the buffer overflows during this interval is smaller than $1 - (1 - \varepsilon')^u < u \cdot \varepsilon'$. Now, recall that if the buffer does not overflow during $[t_0, t_0 + n)$, the buffer will not overflow after time $t_0 + n$. Thus the probability that the buffer will not overflow during an arbitrary busy period is less than $n\varepsilon'$. Finally, let $\varepsilon = n\cdot\varepsilon'$, and apply the result of Lemma 9 for ε', i.e.,

$$s \geq \sqrt{n\left(\frac{\ln n}{2} - \frac{\ln(\varepsilon/n)}{2} - 1\right)} = \sqrt{n\left(\ln n - \frac{\ln \varepsilon}{2} - 1\right)}. \quad \text{(B.45)}$$

□

B.3.2 Arbitrary Flow Rates

In this section we determine the buffer bound for a system in which packets are of unit size, but the reservations can be arbitrary. The basic idea is to use a succession of transformations to reduce the problem to the case in which the probabilities associated to the flows can take, at most, three distinct values, and then to apply the results from the previous case when all reservations are assumed to be identical.

Consider n flows, and let r_k denote the rate reserved by flow k, where

$$\sum_{k=1}^{n} r_k = 1. \quad \text{(B.46)}$$

Consider again the case when all flows are continuously backlogged. Let t_0 denote the starting time of a busy period. Since the time when flow k becomes backlogged is assumed to be independent of t_0, it follows that during the interval $[t_0, t_0+d)$ flow k receives exactly $\lfloor d\cdot r_k\rfloor + 1$ packets with probability

$$p_k(d) = d\cdot r_k - \lfloor d\cdot r_k\rfloor, \quad \text{(B.47)}$$

and $\lfloor d\cdot r_k\rfloor$ packets with probability $1 - p_k(d)$.

Let $p(m;d)$ denote the probability with which the server receives exactly $\sum_{k=1}^{n}\lfloor d\cdot r_k\rfloor + m$ packets during the interval $[t_0, t_0 + d)$. Then

$$p(m;d) = T_n^m(p_1(d), p_2(d), \ldots, p_n(d)), \quad \text{(B.48)}$$

where $T_n^m(p_1(d), p_2(d), \ldots, p_n(d))$ is the coefficient of x^m in the expansion of

$$\prod_{i=1}^{n}(xp_i(d) + (1 - p_i(d))). \quad \text{(B.49)}$$

Note that when all flows have equal reservations, i.e., $r_k = 1/n, 1 \leq k \leq n$, Eq. (B.48) reduces to Eq. (B.28).

By using Eq. (B.47) the number of packets received during $[t_0, t_0+d)$ can be written as

$$\sum_{k=1}^{n}\lfloor d\cdot r_k\rfloor + m = \sum_{k=1}^{n}(d\cdot r_k - p_k(d)) + m = d - \sum_{k=1}^{n}p_k(d) + m. \quad \text{(B.50)}$$

Since t_0 is the starting time of the busy period and since the server remains busy during $[t_0, t_0 + d)$, from Eq. (B.26) it follows that $q(t_0 + d) = m - \sum_{k=1}^{n} p_k(d)$.

Similarly, the probability $P(x > s, u)$ will overflow a queue of size s at time $t_0 + u$ is

$$P(x > s, u) = \sum_{i=v+1}^{n} p(i; u), \tag{B.51}$$

where $v = \sum_{k=1}^{n} p_k(u) + s$.

Since in the following, $p_k(u)$ is always defined over $[t_0, t_0 + u)$ we will drop the argument from the $p_k(u)$'s notation. Next, note that for any two flows k and l, $p(m; u)$ can be rewritten as

$$p(m; u) = p_k p_l A_{k,l}(m) + \tag{B.52}$$
$$(p_k(1 - p_l) + (1 - p_k)p_l)B_{k,l}(m) + (1 - p_k)(1 - p_l)C_{k,l}(m),$$

where $p_k p_l A_{k,l}$, represents all terms in $T_n^m(p_1, p_2, \ldots, p_n)$ that contain $p_k p_l$, $(p_k(1 - p_l) + (1 - p_k)p_l)B_{k,l}$ represents all terms that contain either $p_k(1 - p_l)$ or $(1 - p_k)p_l$, and $(1 - p_k)(1 - p_l)C_{k,l}$ represents all terms that contain $(1 - p_k)(1 - p_l)$.

From Eqs. (B.51) and (B.52), the probability a queue of size s will overflow at time $t_0 + u$ is then

$$P(x > s, u) = \sum_{i=v+1}^{n} p(i; u) \tag{B.53}$$
$$= p_k p_l \cdot \mathcal{A}_{k,l}(v, n) + (p_k(1 - p_l) + (1 - p_k)p_l) \cdot \mathcal{B}_{k,l}(v, n) +$$
$$(1 - p_k)(1 - p_l) \cdot \mathcal{C}_{k,l}(v, n),$$

where $\mathcal{A}_{k,l}(v, n) = \sum_{i=v+1}^{n} A_{k,l}(i)$, $\mathcal{B}_{k,l}(v, n) = \sum_{i=v+1}^{n} B_{k,l}(i)$, and $\mathcal{C}_{k,l}(v, n) = \sum_{i=v+1}^{n} C_{k,l}(i)$, respectively.

Our next goal is to reduce the problem of bounding $P(x > s, u)$ to the case in which the flows' probabilities take a limited number of values. This makes it possible to use the results from the homogeneous reservations case without compromising the bound quality too much. The idea is to iteratively modify the values of the flows' probabilities, without decreasing $P(x > s, u)$. In particular, we consider the following simple transformation: select two probabilities p_k and p_l and update them as follows:

$$p'_k = p_k - \delta, \tag{B.54}$$
$$p'_l = p_l + \delta,$$

where δ is a real value such that $0 \leq p'_l, p'_k \leq 1$, and the new computed probability

$$P'(x > s, u) = p'_k p'_l \cdot \mathcal{A}_{k,l}(v, n) + (p'_k(1 - p'_l) + \tag{B.55}$$
$$(1 - p'_k)p'_l) \cdot \mathcal{B}_{k,l}(v, n) + (1 - p'_k)(1 - p'_l) \cdot \mathcal{C}_{k,l}(v, n).$$

is greater or equal to $P(x > s, u)$.

It is interesting note that performing transformation (B.54) is equivalent to defining a new system in which the reservations of flows k and l are changed to r'_k and r'_l, respectively, such that $p'_k = d \cdot r'_k - \lfloor d \cdot r'_k \rfloor$, and $p'_l = d \cdot r'_l - \lfloor d \cdot r'_l \rfloor$. There are two observations worth noting about this system. First, by choosing $r'_k = r_k - \delta/d$ and $r'_l = r_l + \delta/d$, we maintain the invariant $\sum_{i=1}^n r_i = 1$. Second, while in the new system the start time t_0 of the busy period may change, this will not influence $P'(x > s, u)$ as this depends only on the length of the interval $[t_0, t_0 + u)$.

Next, we give the details of our transformation. From Eqs. (B.53), (B.54) and (B.55), after some simple algebra, we obtain

$$P'(x > s, u) - P(x > s, u) = \delta(p_k - p_l - \delta)\mathcal{D}_{k,l}(v, n), \qquad (B.56)$$

where
$$\mathcal{D}(i, j) = \mathcal{A}_{k,l}(i, j) - 2\mathcal{B}_{k,l}(i, j) + \mathcal{C}_{k,l}(i, j)). \qquad (B.57)$$

Recall that our goal is to choose δ such that $P'(x > s, u) \geq P(x > s)$. Without loss of generality assume that $p_k > p_l$. We consider two cases: (1) if $\mathcal{D}_{k,l}(v, n) > 0$, then $\delta \geq 0$ and $p_k \geq p_l + \delta$ ($\delta < 0$ and $p_k < p_l + \delta$ cannot be simultaneously true); (2) if $\mathcal{D}_{k,l}(v, n) \leq 0$, then either $\delta \geq 0$ and $p_k \leq p_l + \delta$, or $\delta < 0$ and $p_k > p_l + \delta$.

Let $p_{min} = \min_{1 \leq i \leq n} p_i$, and $p_{max} = \max_{1 \leq i \leq n} p_i$, respectively. Consider the following three subsets, denoted U, V, and M, where U contains all flows k such that $p_k = p_{min}$, V contains all flows k such that $p_k = p_{max}$, and M contains all the other flows. The idea is then to successively apply the transformation (B.54) on $p_1, p_2, \ldots, p_n$, until the probabilities of all flows in M become equal. In this way we reduce the problem to the case in which the probabilities p_k can take at most three distinct values: p_{min}, p_{max}, and p_M, where $p_k = p_M$, $\forall k \in M$. Figure B.2 shows the iterative algorithm that achieves this. Lemmas 10 and 11 prove that by using the algorithm in Figure B.2, $p_1, p_2, \ldots, p_n$ converge asymptotically to the three values.

while $(|M| > 1)$ **do** /* *while size of M is greater than one* */
    $p_l = \min_{i \in M}(p_i)$;
    $p_k = \max_{i \in M}(p_i)$;
    if $(D_{k,l}(v, n) > 0)$
        $p_k = p_l = (p_k + p_l)/2$;
    else
        $\delta = \max(p_k - p_{max}, p_{min} - p_l)$;
        $p_k = p_k - \delta$; $p_l = p_l + \delta$;
        if $(p_l = p_{min})$
            $M = M \setminus \{l\}$; $U = U \cup \{l\}$;
        if $(p_k = p_{max})$
            $M = M \setminus \{k\}$; $V = V \cup \{k\}$;

Fig. B.2. Reducing $p_1, p_2, \ldots p_n$ to three distinct values.

Lemma 10 *After an iteration of the algorithm in Figure B.2, either the size of M decreases by one, or the standard deviation of the probabilities in M decreases by a factor of at least $(1 - \frac{1}{2|M|})$.*

Proof. The first part is trivial; if $D_{k,l}(v, n) \leq 0$ the size of M decreases by one. For the second part, let $\bar{p}$ denote the average values of probabilities associated to the flows in M, i.e.,

$$\bar{p} = \frac{\sum_{i \in M} p_i}{|M|}. \tag{B.58}$$

The standard deviation associated to the probabilities in M is

$$dev = \sum_{i \in M} (p_i - \bar{p})^2. \tag{B.59}$$

After averaging probabilities p_k and p_l, standard deviation v changes to

$$dev' = dev + 2 \left(\frac{p_k + p_l}{2} - \bar{p} \right)^2 - (p_k - \bar{p})^2 - (p_l - \bar{p})^2 \tag{B.60}$$

$$= dev - \frac{(p_k - p_l)^2}{2}.$$

Since p_k and p_l are the lowest, and respectively, the highest probabilities in M we have $(p_i - \bar{p})^2 \leq (p_l - p_k)^2$, $\forall i \in M$. From here and from Eqs. (B.59) and (B.60) we have

$$dev = \sum_{i \in M} (p_i - \bar{p})^2 \leq |M|(p_l - p_k)^2 = 2|M|(dev - dev') \Rightarrow \tag{B.61}$$

$$dev' \leq dev \cdot \left(1 - \frac{1}{2|M|} \right).$$

□

Lemma 11 *Consider n flows, and let p_i denote the probability associated with flow i. Then, by using the algorithm in Figure B.2, the probabilities p_i $(1 \leq i \leq n)$ converge to, at most, three values.*

Proof. Let ε be an arbitrary small real. The idea is then to show that after a finite number of iterations of the algorithm in Figure B.2, the standard deviation of p_i's $(i \in M)$ becomes smaller than ε.

The standard deviation for the probabilities of flows in M is trivially bounded as follows

$$dev = \sum_{i \in M} (p_i - \bar{p})^2 \leq \sum_{i \in M} (p_{max} - p_{min})^2 \tag{B.62}$$

$$= |M|(p_{max} - p_{min})^2 < n(p_{max} - p_{min})^2.$$

Assume $D_{k,l}(v, n) > 0$ (i.e., M does not change) for n_1 consecutive iterations. Then, by using Lemma 10, it is easy to see that n_1 is bounded above by N, where

$$dev \cdot \left(1 - \frac{1}{2|M|}\right)^N < dev \cdot \left(1 - \frac{1}{2n}\right)^N = \varepsilon \Rightarrow N = \frac{\ln(\varepsilon/dev)}{\ln(1 - 1/(2n))} \quad \text{(B.63)}$$

Since the above bound, N, holds for any set M, it follows that after nN iterations, we are guaranteed that either set M becomes empty, a case in which the lemma is trivially true, or $dev < \varepsilon$. $\square$

Thus, we have reduced the problem to compute an upper bound for probability $P(x > s, u)$ in a system in which probabilities take only three values at time u: p_{min}, p_{max}, and p_M.

Next we give the main result of this section

Lemma 12 *Consider n flows with unit packet sizes and arbitrary flow reservations. Then given a buffer of size s, where*

$$s \geq \sqrt{3n\left(\frac{\ln n}{2} - \frac{\ln \varepsilon}{2} - 1\right)}, \quad \text{(B.64)}$$

the probability that the buffer overflows in an arbitrary time slot during a server busy period is asymptotically $< \varepsilon$.

Proof. Consider the probability, $P(x > s, u)$, with which the queue overflows at time $t_0 + u$ (see Eq. (B.51)). Next, by using the algorithm in Figure B.2, we reduce probabilities p_i's ($1 \leq i \leq n$) to three values: p_{min}, p_{max}, and p_M, respectively. Let p_i^f denote the final probability of flow i, and let $P^f(x > s, u)$ denote the final probability of the queue overflowing at time $t_0 + u$. More precisely, from Eqs. (B.51) and (B.48) we have

$$P^f(x > s, u) = \sum_{i=v+1}^{n} p^f(i; u) = \sum_{i=v+1}^{n} T_n^i(p_1, p_2, \ldots, p_n), \quad \text{(B.65)}$$

where $v = \sum_{k=1}^{n} p_k(u) + s$, and $p_i = p_{min}, \forall i \in U, p_i = p_{max}, \forall i \in V$, and $p_i = p_M, \forall i \in V$. Since after each transformation $P(x > s, u)$ can only increase, we have $P^f(x > s, u) \geq P(x > s, u)$.

Let n_U, n_V, and n_M be the number of flows in sets U, V, and M, respectively. Define integers v_u, v_V, and v_M, such that $v = v_u + v_V + v_M$, and $v_U < n_U, v_V < n_V$, and $v_M < n_M$, respectively. Then, it can be shown that

$$P^f(x > s, u) \leq P_U + P_V + P_M, \quad \text{(B.66)}$$

where

$$P_U = \sum_{i=v_U+1}^{n_U} \binom{n_U}{i} p_{min}^i (1 - p_{min})^{n_U - i} \quad \text{(B.67)}$$

$$P_V = \sum_{i=v_V+1}^{n_V} \binom{n_V}{i} p_{max}^i (1 - p_{max})^{n_V - i}$$

$$P_M = \sum_{i=v_M+1}^{n_M} \binom{n_M}{i} p_M^i (1 - p_M)^{n_M - i}.$$

Due to the notation complexity we omit the derivation of Eq. (B.66). Instead, below we give an alternate method that achieves the same result.

The key observation is that P_U represents the probability with which more than $\sum_{i \in U} \lfloor u \cdot r_i \rfloor + v_U$ packets from flows in U arrive during the interval $[t_0, t_0 + u)$. This is easy to see, as the probability that *exactly* $\sum_{i \in U} \lfloor u \cdot r_i \rfloor + m$ packets from flows in U arrive during $[t_0, t_0 + u)$ is $\binom{n_U}{m} p_{min}^m (1 - p_{min})^{n_U - m}$ (see Eq. (B.48) for comparison).

Similarly, P_V is the probability that more than $\sum_{i \in V} \lfloor u \cdot r_i \rfloor + v_V$ packets from flows in V arrive during $[t_0, t_0 + u)$, while P_M is the probability that more than $\sum_{i \in M} \lfloor u \cdot r_i \rfloor + v_M$ packets from flows in M arrive during the same interval.

Consequently, $(1 - P_U)(1 - P_V)(1 - P_M)$ represents the probability with which *no* more than $\sum_{i \in U} \lfloor u \cdot r_i \rfloor + v_U$, $\sum_{i \in V} \lfloor u \cdot r_i \rfloor + v_V$, and $\sum_{i \in M} \lfloor u \cdot r_i \rfloor + v_M$ packets are received from flows in U, V, and M during $[t_0, t_0 + u)$. Clearly this probability is *no larger* than the probability of receiving no more than $\sum_{i=1}^{n} \lfloor u \cdot r_i \rfloor + v$ packets from *all* flows during the interval $[t_0, t_0 + u)$, a probability which is exactly $1 - P^f(x > s, u)$. This yields

$$1 - P^f(x > s, u) \geq (1 - P_U)(1 - P_V)(1 - P_M) \Rightarrow \qquad (B.68)$$
$$P^f(x > s, u) \leq 1 - (1 - P_U)(1 - P_V)(1 - P_M) \leq P_U + P_V + P_M.$$

Next, consider the expression of P_U in Eq. (B.67). Let

$$s_U = v_U - u_U, \qquad (B.69)$$

where $u_U = p_{min} n_U$. Then it is easy to see that the expressions of p_{min} (i.e., $p_{min} = u_U / n_U$) and P_U, given by Eq. (B.67), are identical to the expressions of $p(d)$ and $P(x > s, u)$, given by Eqs. (B.27) and (B.29), respectively, after the following substitutions: $d \leftarrow u_U$, $n \leftarrow n_U$, $u \leftarrow u_U$, $s \leftarrow s_U$. By applying the result of Lemma 8 we have the following bound

$$P_U = \sum_{i=u_U+s_U+1}^{n_U} \binom{n_U}{i} p_{min}^i (1 - p_{min})^{n_U - i} \qquad (B.70)$$
$$< \beta(n_U) \sqrt{\frac{1}{2\pi}} \left(\frac{1 - (s_U - 1)/2n_U}{1 + (s_U - 1)/2n_U} \right)^{2s_U} \frac{(n_U + s_U)^2}{4 s_U n_U}.$$

Next we compute s_U, such that

$$\frac{\varepsilon}{3} = \beta(n_U) \sqrt{\frac{1}{2\pi}} \left(\frac{1 - (s_U - 1)/2n_U}{1 + (s_U - 1)/2n_U} \right)^{2s_U} \frac{(n_U + s_U)^2}{4 s_U n_U}. \qquad (B.71)$$

By applying the same approximations used in proving Lemma 9 (see Eq. (B.42)), i.e., $s_U \ll n_U$, $s_V \ll n_V$, and $s_M \ll n_M$, respectively, we get

$$s_U \simeq \sqrt{n_U \left(\frac{\ln n_U}{2} - \frac{\ln(\varepsilon/3)}{2} - 1 \right)}, \qquad (B.72)$$

and similarly

$$s_V \simeq \sqrt{n_V \left(\frac{\ln n_V}{2} - \frac{\ln(\varepsilon/3)}{2} - 1 \right)} \tag{B.73}$$

$$s_M \simeq \sqrt{n_M \left(\frac{\ln n_M}{2} - \frac{\ln(\varepsilon/3)}{2} - 1 \right)}.$$

By using the above values for s_U, s_V, and s_M, respectively, and by the definition of $P^f(x > s, u)$ and Eq. (B.66), we have

$$P(x > s, u) \leq P^f(x > s, u) \leq P_U + P_V + P_M \leq 3 \cdot \frac{\varepsilon}{3} = \varepsilon. \tag{B.74}$$

Now it remains to compute s. First, recall that $s_U = v_U - u_U$, $s_V = v_U - u_V$, $s_M = v_M - u_M$, where $u_U = p_{min}u$, $u_V = p_{max}u$, and $u_M = p_M u$ (see Eq. (B.69)). From here we obtain

$$s_U + s_V + s_M = (v_U - n_U) + (v_V - n_V) + (v_M - n_M) \tag{B.75}$$

$$= v - n_U - n_V - n_M - 3$$

$$= v - p_{min}n_U - p_{max}n_V - p_M n_M$$

$$= v - \sum_{i \in U} p_{min} - \sum_{i \in V} p_{max} - \sum_{i \in M} p_M$$

$$= v - \sum_{i=1}^{n} p_i = s.$$

As both $P(x > s, u)$ and $P^f(x > s, u)$ decrease in s, for our purpose it is sufficient to determine an upper bound for s. From Eqs. (B.72), (B.73) and (B.75) this reduces to compute

$$\max \left(\sum_{I \in \{U,V,M\}} \sqrt{n_I \left(\frac{\ln n_I}{2} - \frac{\ln(\varepsilon/3)}{2} - 1 \right)} \right), \tag{B.76}$$

subject to $n_U + n_V + n_M = n$. Since the function $\sqrt{x \ln x}$ is concave, it follows that expression (B.76) achieves maximum for $n_U = n_V = n_M = n/3$. Finally, we choose

$$s = 3 \cdot \sqrt{\frac{n}{3} \left(\frac{\ln(n/3)}{2} - \frac{\ln(\varepsilon/3)}{2} - 1 \right)} = \sqrt{3n \left(\frac{\ln n}{2} - \frac{\ln \varepsilon}{2} - 1 \right)}, \tag{B.77}$$

which completes the proof. $\square$

By combining Lemmas 9 and 12 we have the following result

Theorem 3 *Consider a server traversed by n flows. Assume that the arrival times of the packets from different flows are independent, and that all packets have the same size. Then, for any given probability ε, the queue size at any* *time instant during a server busy period is asymptotically bounded above by s, where*

$$s = \sqrt{\beta n \left(\frac{\ln n}{2} - \frac{\ln \varepsilon}{2} - 1 \right)}, \tag{B.78}$$

with a probability larger than $1 - \varepsilon$. For identical reservations $\beta = 1$; for heterogeneous reservations $\beta = 3$.

B.4 Proof of Theorem 4

Theorem 4 *Consider a link of capacity C at time t. Assume that no reservation terminates and there are no reservation failures or request losses after time t. Then if there is sufficient demand after t the link utilization approaches asymptotically $C(1 - f)/(1 + f)$.*

Proof. If the aggregate reservation at time t is larger than $C(1 - f)/(1 + f)$, the proof is trivially true. Next, we consider the case in which the aggregate reservation is less than $C(1 - f)/(1 + f)$.

In particular, let $C(1 - f)/(1 + f) - \Delta$ be the aggregate reservation at time t. Without loss of generality assume $t = u_k$. Then we will show that if no reservation terminates, no reservation request fails, and there is enough demand after time u_k, then at least $(1 + f)\Delta/2$ bandwidth is allocated during the next two slots. i.e., during the interval $(u_k, u_{k+2}]$. Thus, for any arbitrary small real ε, we are guaranteed that after, at most,

$$2 \times \frac{\ln(\varepsilon/\Delta)}{\ln((1 - f)/2)} \tag{B.79}$$

slots the aggregate reservation will exceed $C(1 - f)/(1 + f) - \varepsilon$.

From Eq. (5.20) it follows that the maximum capacity which can be allocated during the interval $(u_k, u_{k+1}]$ is $\max(C - R_{cal}(u_k), 0)$. Assume then that Δ_1 capacity is allocated during $(u_k, u_{k+1}]$, where $\Delta_1 \leq \max(C - R_{cal}(u_k), 0)$. Consider two cases whether $\Delta_1 \geq \Delta$ or not. If $\Delta_1 \geq \Delta$, the proof follows trivially.

Assume $\Delta_1 < \Delta$. Then we will show that at time u_{k+2} the aggregate reservation can increase by at least a constant fraction of Δ. From Figure B.3 is easy to see that, for any reservation continuously active during an interval $(u_k, u_{k+1}]$, we have

$$b_i(u_k, u_{k+1}) < r_i(T_W + T_I + T_J). \tag{B.80}$$

Since no reservation terminates during $(u_k, u_{k+1}]$, we have $\mathcal{L}(u_{k+1}) = \mathcal{L}(u_k) \cup \mathcal{N}(u_{k+1})$. Let $ac_i \in (u_k, u_{k+1}]$ be the time when flow i becomes active during $(u_k, u_{k+1}]$. Since $b_i(ac_i, u_{k+1}) \leq b_i(u_k, u_{k+1})$, by using Eq. (B.80), we obtain

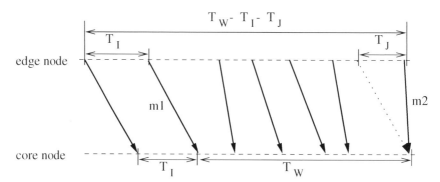

Fig. B.3. The scenario in which the upper bound of b_i, i.e., $r_i(T_W - T_I - T_J)$, is achieved. The arrows represent packet transmissions. T_W is the averaging window size; T_I is an upper bound on the packet inter-departure time; T_J is an upper bound on the delay jitter. Both $m1$ and $m2$ fall just inside the estimation interval, T_W, at the core node.

$$B(u_k, u_{k+1}) = \sum_{i \in \mathcal{L}(u_{k+1})} b_i(u_k, u_{k+1}) < \sum_{i \in \mathcal{L}(u_{k+1})} r_i(T_W + T_I + T_W). \quad \text{(B.81)}$$

From here we get

$$R_{DPS}(u_k, u_{k+1}) < R(u_{k+1})(1 + f). \quad \text{(B.82)}$$

Since there are no duplicate requests or partial reservation failures after time $t = u_k$, we have $\Delta_1 = R_{new}(u_{k+1})$. From here and from Eq. (5.20) and Eq. (B.82) we have

$$R_{cal}(u_{k+1}) \leq \frac{R_{DPS}(u_{k+1})}{1 - f} + \Delta_1 < R(u_{k+1})\frac{1 + f}{1 - f} + \Delta_1. \quad \text{(B.83)}$$

In addition, we have $R(u_{k+1}) = R(u_k) + \Delta_1$. Since $R(u_k) = C(1 - f)/(1 + f) - \Delta$, from Eq. (B.83), it follows

$$C - R_{cal}(u_{k+1}) \geq C - R(u_{k+1})\frac{1 + f}{1 - f} - \Delta_1 \geq \frac{1 + f}{1 - f}\Delta - \frac{2}{1 - f}\Delta_1. \quad \text{(B.84)}$$

Finally, consider two cases whether (a) $\Delta_1 < \Delta(1 + f)/2$, or (b) not. If (a) is true then the link can allocate up to

$$\Delta_1 + C - R_{cal}(u_{k+1}) > \Delta_1 + \frac{1 + f}{1 - f}\Delta - \frac{2}{1 - f}\Delta_1 = \frac{1 + f}{1 - f}(\Delta - \Delta_1) > \frac{1 + f}{2}\Delta,$$
$$\text{(B.85)}$$

capacity during the time interval $(u_k, u_{k+2}]$. In case (b) we have trivially $\Delta_1 \geq \Delta(1 + f)/2$. Thus in both cases we can allocate at least $\Delta(1 + f)/2$ new capacity during $(u_k, u_{k+2}]$. $\square$

References

1. ATM User-network Interface (UNI) signalling specification version 4.0, July 1996. The ATM Forum Technical Committee, af-sig-0061.000.
2. Avici terabit switch router. Product Brochure, http://www.avici.com/products/index.html.
3. R. Y. Awdeh and H. T. Mouftah. Survey of ATM switch architectures. *Computer Networks and ISDN Systems*, pages 1567–1613, September 1995.
4. Ö. Babaoğlu and S. Toueg. Non-blocking atomic commitment. *Distributed Systems, S. Mullender (ed.)*, pages 147–168, 1993.
5. F. Baker, C. Iturralde, F. Le Faucheur, and B. Davie. Aggregation of RSVP for IP4 and IP6 reservations. Internet Draft, draft-baker-rsvp-aggregation-00.txt.
6. H. Balakrishnan, H. Rahul, and S. Seshan. An integrated congestion management architecture for Internet hosts. In *Proceedings of ACM SIGCOMM'99*, pages 175–188, Cambridge, MA, September 1999.
7. M. Baldi and Y. Ofek. End-to-end delay analysis of videoconferencing over packet switched networks. *IEEE/ACM Transactions on Networking*, 4(8):479–492, August 2000.
8. A. Banerjea and B. Mah. The real-time channel administration protocol. In *Proceedings of NOSSDAV'91*, pages 160–170, Heidelberg, Germany, November 1991. Springer-Verlag.
9. J.C.R. Bennett, D.C. Stephens, and H. Zhang. High speed, scalable, and accurate implementation of packet fair queueing algorithms in ATM networks. In *Proceedings of IEEE ICNP '97*, pages 7–14, Atlanta, GA, October 1997.
10. J.C.R. Bennett and H. Zhang. WF^{2}Q: Worst-case fair weighted fair queueing. In *Proceedings of IEEE INFOCOM'96*, pages 120–128, San Francisco, CA, March 1996.
11. J.C.R. Bennett and H. Zhang. Hierarchical packet fair queueing algorithms. *IEEE/ACM Transactions on Networking*, 5(5):675–689, October 1997.
12. U. Black. *ATM: Foundation for Broadband Networks*. Prentice Hall, 1995.
13. S. Blake, D. Black, M. Carlson, E. Davies, Z. Wang, and W. Weiss. An architecture for differentiated services, December 1998. Internet RFC 2475.
14. L. Breslau, E. W. Knightly, S. Shenker, I. Stoica, and H. Zhang. Endpoint admission control: Architectural issues and performance. In *Proceedings of ACM SIGCOMM'00*, pages 57–69, Stockholm, Sweden, September 2000.
15. R. Brown. Calendar queues: A fast O(1) priority queue implementation for the simulation event set problem. *Communications of the ACM*, 31(10):1220–1227, October 1988.
16. Collaborative advanced interagency research network (cairn). http://www.cairn.net/.

17. R. Callon, P. Doolan, N. Feldman, A. Fredette, G. Swallow, and A. Viswanathan. A framework for multiprotocol label switching, November 1997. Internet Draft, draft-ietf-mpls-framework-02.txt.

18. Z. Cao, Z. Wang, and E. Zegura. Rainbow fair queueing: Fair bandwidth sharing without per-flow state. In *Proceedings of INFOCOM'99*, pages 922–931, Tel-Aviv, Israel, March 2000.

19. P. Chandra, A. Fisher, C. Kosak, T. S. E. Ng, P. Steenkiste, E. Takahashi, and H. Zhan. Darwin: Resource management for value-added customizable network services. In *Proceedings of IEEE ICNP'98*, pages 177,196, AUSTIN, TX, October 1998.

20. A. Charny. An algorithm for rate allocation in a packet-switching network with feedback. Master's thesis, MIT, CS Division, May 1994.

21. S. T. Chuang, A. Goel, N. McKeown, and B. Prabhakar. Matching output queueing with a combined input output queued switch. In *Proceedings of INFOCOM'99*, pages 1169–1178, New York, CA, March 1999.

22. D. Clark. The design philosophy of the DARPA internet protocols. In *Proceedings of ACM SIGCOMM'88*, pages 106–114, Stanford, CA, August 1988.

23. D. Clark. Internet cost allocation and pricing. *Internet Economics, L. W. McKnight and J. P. Bailey (eds.)*, pages 215–252, 1997.

24. D. Clark and J. Wroclawski. An approach to service allocation in the Internet, July 1997. Internet Draft, http://diffserv.lcs.mit.edu/draft-clark-diff-svc-alloc-00.txt.

25. A. Clerget and W. Dabbous. Tag-based unified fairness. In *Proceedings of INFOCOM'01*, pages 498–507, Anchorage, AK, April 2001.

26. T. H. Cormen, C. E. Leiserson, and R. L. Rivest. *Introduction to Algorithms*. The MIT Press, 1990.

27. M. E. Crovella and A. Bestavros. Self-similarity in world wide web traffic evidence and possible causes. In *Proceedings of ACM SIGMETRICS 96*, pages 160–169, Philadelphia, PA, May 1996.

28. R. Cruz. Quality of service guarantees in virtual circuit switched network. *IEEE Journal of Selected Area on Communications*, 13(6):1048–1056, August 1995.

29. R. L. Cruz. SCED+: Efficient management of quality of service guarantees. In *Proceedings of INFOCOM'98*, pages 625–642, San Francisco, CA, 1998.

30. M. Degermark, A. Brodnik, S. Carlsson, and S. Pink. Small forwarding tables for fast routing lookups. In *Proceedings of ACM SIGCOMM'97*, pages 3–14, Cannes, France, September 1997.

31. A. Demers, S. Keshav, and S. Shenker. Analysis and simulation of a fair queueing algorithm. In *Journal of Internetworking Research and Experience*, pages 3–26, October 1990.

32. Y. Bernet et. al. A framework for differentiated services, November 1998. Internet Draft, draft-ietf-diffserv-framework-01.txt.

33. S. Fahmy, R. Jain, S. Kalyanaraman, R. Goyal, and B. Vandalore. On determining the fair bandwidth share for ABR connections in ATM networks. In *Proceedings of IEEE ICC '98*, volume 3, pages 1485–1491, Atlanta, GA, June 1998.

34. D. Ferrari and D. Verma. A scheme for real-time channel establishment in wide-area networks. *IEEE Journal of Selected Area on Communications*, 8(3):368–379, April 1990.

35. N. Figueira and J. Pasquale. An upper bound on delay for the VirtualClock service discipline. *IEEE/ACM Transactions on Networking*, 3(4):399,408, August 1995.

36. S. Floyd and K. Fall. Promoting the use of end-to-end congestion control in the internet. *IEEE/ACM Transactions on Networking*, 7(4):458–472, August 1999.

37. S. Floyd and V. Jacobson. Random early detection for congestion avoidance. *IEEE/ACM Transactions on Networking*, 1(4):397–413, July 1993.

38. S. Floyd and V. Jacobson. The synchronization of periodic routing messages. In *Proceedings of ACM SIGCOMM'93*, pages 33–44, San Francisco, CA, September 1993.

39. S. Floyd and V. Jacobson. Link-sharing and resource management models for packet networks. *IEEE/ACM Transactions on Networking*, 3(4):365–386, August 1995.

40. R. Frederick. Network video (nv). Software available via ftp://ftp.parc.xerox.com/net-research.

41. V. Fuller, T. Li, J. Yu, and K. Varadhan. Classless inter-domain routing (CIDR): An address assignment and aggregation strategy, September 1993.

42. L. Georgiadis, R. Guerin, V. Peris, and K. Sivarajan. Efficient network QoS provisioning based on per node traffic shaping. *IEEE/ACM Transactions on Networking*, 4(4):482–501, August 1996.

43. D. W. Glazer and C. Tropper. A new metric for dynamic routing algorithms. *IEEE Transaction on Communication*, 38(3):360–367, March 1990.

44. S. Golestani. A stop-and-go queueing framework for congestion management. In *Proceedings of ACM SIGCOMM'90*, pages 8–18, Philadelphia, PA, September 1990.

45. S. Golestani. A self-clocked fair queueing scheme for broadband applications. In *Proceedings of IEEE INFOCOM'94*, pages 636–646, Toronto, CA, June 1994.

46. S. J. Golestani. *A Unified Theory of Flow Control and Routing in Communication Networks*. PhD thesis, MIT, Department of EECS, May 1980.

47. P. Goyal, S. Lam, and H. Vin. Determining end-to-end delay bounds in heterogeneous networks. In *Proceedings of NOSSDAV'95*, pages 287–298, Durham, New Hampshire, April 1995.

48. P. Goyal, H.M. Vin, and H. Chen. Start-time Fair Queuing: A scheduling algorithm for integrated services. In *Proceedings of ACM SIGCOMM 96*, pages 157–168, Palo Alto, CA, August 1996.

49. R. Guerin, S. Blake, and S. Herzog. Aggregating RSVP-based QoS requests, November 1997. Internet Draft, draft-guerin-aggreg-rsvp-00.txt.

50. P. Gupta, S. Lin, and N. McKeown. Routing lookups in hardware at memory access speeds. In *Proceedings of INFOCOM'98*, pages 1240–1256, San Francisco, CA, 1998.

51. Pankaj Gupta and Nick McKeown. Packet classification on multiple fields. In *Proceedings of ACM SIGCOMM'99*, pages 147–160, Cambridge, MA, September 1999.

52. E.L. Hahne. Round-robin scheduling for max-min fairness in data networks. *IEEE Journal of Selected Area on Communications*, 9(7):1024–1039, September 1991.

53. G. Hardin. The tragedy of the commons. *Science*, 162:1243–1248, 1968.

54. C. Hedrick. Routing information protocol, June 1988. Internet RFC 1058.
55. Juha Heinanen, F. Baker, W. Weiss, and J. Wroclawski. Assured forwarding PHB group, June 1999. Internet RFC 2597.
56. Internet assignment number authority (iana). IP Option Numbers, http://www.isi.edu/in-notes/iana/assignments/ip-parameters.
57. V. Jacobson. Congestion avoidance and control. In *Proceedings of ACM SIGCOMM'88*, pages 314–329, August 1988.
58. V. Jacobson and S. McCanne. LBL whiteboard (wb). Software available via ftp://ftp.ee.lbl.gov/conferencing/wb.
59. V. Jacobson and S. McCanne. Visual audio tool (vat). Software available via ftp://ftp.ee.lbl.gov/conferencing/vat.
60. Van Jacobson and K. Poduri K. Nichols. An expedited forwarding PHB, June 1999. Internet RFC 2598.
61. R. Jain, S. Kalyanaraman, R. Goyal, S. Fahmy, and R. Viswanathan. ERICA switch algorithm: A complete description, August 1996. ATM Forum/96-1172.
62. S. Jamin, P. Danzig, S. Shenker, and L. Zhang. A measurement-based admission control algorithm for integrated services packet networks. In *Proceedings of SIGCOMM'95*, pages 2–13, Boston, MA, September 1995.
63. C. Kalmanek, H. Kanakia, and S. Keshav. Rate controlled servers for very high-speed networks. In *Proceedings of IEEE Globecom*, pages 300.3.1 – 300.3.9, San Diego, California, December 1990.
64. G. Karlsson. Asynchronous transfer of video. *IEEE Communication Magazine*, pages 118–126, August 1996.
65. L. Kleinrock. *Queueing Systems*. John Wiley and Sons, 1975.
66. T.V. Lakshman and D. Stiliadis. High speed policy-based packet forwarding using efficient multi-dimensional range matching. In *Proceedings of ACM SIGCOMM'98*, pages 203–214, Vancouver, Canada, September 1998.
67. D. Lin and R. Morris. Dynamics of random early detection. In *Proceedings of ACM SIGCOMM'97*, pages 127–137, Cannes, France, October 1997.
68. Q. Ma, P. Steenkiste, and H. Zhang. Routing high-bandwidth traffic in max-min fair share networks. In *Proceedings of ACM SIGCOMM'96*, pages 206–217, Palo Alto, CA, October 1996.
69. S. McCanne. *Scalable Compression and Transmission of Internet Multicast Video*. PhD thesis, University of California at Berkeley, Computer Science Division, December 1996. Technical Report UCB/CSD-96-928.
70. S. McCanne and V. Jacobson. vic: A flexible framework for packet video. In *Proceedings of ACM Multimedia'95*, pages 511–522, San Francisco, CA, November 1995.
71. J. M. McQuillan, I. Richer, and E. Rosen. The new routing algorithm for the arpanet. *IEEE Transaction on Communication*, 28(5):711–719, May 1980.
72. D. R. Morrison. Patricia — practical algorithms to retrieve information coded in alphanumeric. *Journal of the ACM*, pages 514–534, October 1968.
73. J. Moy. OSPF version 2, July 1991. Internet RFC 1247.
74. J Nagle. On packet switches with infinite storage. *IEEE Transactions On Communications*, 35(4):435–438, April 1987.
75. K. Nichols, S. Blake, F. Baker, and D. L. Black. Definition of the differentiated services field (DS field) in the IPv4 and IPv6 headers, December 1998. Internet RFC 2474.
76. K. Nichols, V. Jacobson, and L. Zhang. An approach to service allocation in the Internet, November 1997. Internet Draft.

77. Network traffic packet header traces. URL: http://moat.nlanr.net/Traces/.
78. Ucb/lbnl/vint network simulator - ns (version 2). `http://www-mash.cs.berkeley.edu/ns/`.
79. A. Parekh and R. Gallager. A generalized processor sharing approach to flow control - the single node case. *ACM/IEEE Transactions on Networking*, 1(3):344–357, June 1993.
80. V. Paxon and S. Floyd. Wide-area traffic: The failure of poisson modeling. *IEEE/ACM Transactions on Networking*, 3(3):226–244, June 1995.
81. V. Paxon and S. Floyd. Why we don't know how to simulate the Internet. In *Proceedings of the Winder Communication Conference*, December 1997. http://ftp.ee.lbl.gov/nrg-papers.html.
82. S. Shenker R. Braden, D. Clark. Integrated services in the Internet architecture: An overview, June 1994. Internet RFC 1633.
83. K. Ramakrishnan, D. Chiu, and R. Jain. Congestion avoidance in computer networks with a connectionless network layer. In *Proceedings of ACM SIG-COMM'88*, pages 303–313, Stanford, CA, August 1988.
84. R. Rejaie, M. Handley, and D. Estrin. Quality adaptation for congestion control video playback over the Internet. In *Proceedings of ACM SIGCOMM'99*, pages 189–200, Cambridge, MA, September 1999.
85. R. Rejaie, M. Handley, and D. Estrin. RAP: An end-to-end rate-based congestion control mechanism for realtime streams in the Internet. In *Proceedings of INFOCOM'99*, pages 1337–1345, New York, NY, 1999.
86. Y. Rekhter and T. Li. A border gateway protocol 4 (BGP-4), March 1995. Internet RFC 1771.
87. L. Roberts. Enhanced PRCA (proportional rate control algorithm), August 1994. ATM Forum/94-0735R1.
88. L. G. Roberts. Internet growth trends. http://www.ziplink.net/
89. T. Rutkowski. Bianual strategic note. *Center for Next Generation Internet*, February 2000. http://www.ngi.org/trends/TrendsPR0002.txt.
90. H. Sariowan, R.L. Cruz, and G.C. Polyzos. Scheduling for quality of service guarantees via service curves. In *Proceedings of the International Conference on Computer Communications and Networks (ICCCN) 1995*, pages 512–520, September 1995.
91. S. Savage, D. Wetherall, A. Karlin, and T. Anderson. Practical network support for IP traceback. In *Proceedings of ACM SIGCOMM'00*, pages 295–306, Stockholm, Sweden, September 2000.
92. S. Shenker, D. C. Clark, and L. Zhang. A scheduling service model and a scheduling architecture for an integrated services packet network, 1993. ftp://parcftp.parc.xerox.com/pub/net-research/archfin.ps.
93. S. Shenker, C. Partridge, and R. Guerin. Specification of guaranteed quality of service, September 1997. Internet RFC 2212.
94. M. Shreedhar and G. Varghese. Efficient fair queueing using deficit round robin. In *Proceedings of SIGCOMM'95*, pages 231–243, Boston, MA, September 1995.
95. K. Y. Siu and H. Y. Tzeng. Intelligent congestion control for ABR service in ATM networks, July 1994. Technical Report No. 1102, ECE, UC Irvine.
96. V. Srinivasan, S. Suri, and G. Varghese. Packet classification using tuple space search. In *Proceedings of ACM SIGCOMM'99*, pages 135–146, Cambridge, MA, September 1999.

97. V. Srinivasan, G. Varghese, S. Suri, and M. Waldvogel. Fast scalable algorithms for level four switching. In *Proceedings of ACM SIGCOMM'98*, pages 191–202, Vancouver, Canada, September 1998.

98. D.C. Stephens, J.C.R. Bennett, and H.Zhang. Implementing scheduling algorithms in high speed networks. *IEEE Journal of Selected Area on Communications: Special Issue on Next-generation IP Switches and Router*, 17(6):1145–1158, June 1999.

99. D. Stilliadis and A. Varma. Efficient fair queueing algorithms for packet-switched networks. *IEEE/ACM Transactions on Networking*, 6(2):175–185, April 1998.

100. D. Stilliadis and A. Verma. Latency-rate servers: A general model for analysis of traffic scheduling algorithms. *IEEE/ACM Transactions on Networking*, 6(2):164–174, April 1998.

101. I. Stoica, S. Shenker, and H. Zhang. Core-stateless fair queueing: Achieving approximately fair bandwidth allocations in high speed networks. In *Proceedings ACM SIGCOMM'98*, pages 118–130, Vancouver, September 1998.

102. I. Stoica and H. Zhang. Exact emulation of an output queueing switch by a combined input output queueing switch. In *Proceedings of IWQoS'98*, pages 218–224, Napa, CA, 1998.

103. I. Stoica and H. Zhang. LIRA: A model for service differentiation in the Internet. In *Proceedings of NOSSDAV'98*, pages 115,128, London, UK, July 1998.

104. I. Stoica and H. Zhang. Providing guaranteed services without per flow management. In *Proceedings of ACM SIGCOMM'99*, pages 81–94, Cambridge, MA, September 1999.

105. I. Stoica and H. Zhang. Providing guranteed services without per flow management, May 1999. Technical Report CMU-CS-99-133.

106. I. Stoica, H. Zhang, and T.S.E. Ng. A hierarchical fair service curve algorithm for link-sharing, real-time and priority service. In *Proceedings of ACM SIGCOMM'97*, pages 162–173, Cannes, Frances, September 1997.

107. I. Stoica, H. Zhang, S. Shenker, R. Yavatkar, D. Stephens, Y. Bernet, Z. Wang, F. Baker, J. Wroclawski, and R. Wilder C. Song. Per hop behaviors based on dynamic packet states, February 1999. Internet Draft, draft-stoica-diffserv-dps-00.txt.

108. B. Suter, T.V. Lakshman, D. Stiliadis, and A.K. Choudhury. Buffer management schemes for supporting TCP in gigabit routers with per-flow queueing. *IEEE Journal on Selected Areas in Communication*, August 1999.

109. A. Tanenebaum. *Computer Networks*. Prentice Hall, 1996.

110. D. Tse and M. Grosslauser. Measurement-based call admission control: Analysis and simulation. In *Proceedings of INFOCOM'97*, pages 981–989, Kobe, Japan, 1997.

111. N. Venkitaraman, J. Mysore, R. Srikant, and R. Barnes. Stateless prioritized fair queuing, August 2000. Internet Draft, draft-venkitaraman-diffserv-spfq-00.txt.

112. D. Verma, H. Zhang, and D. Ferrari. Guaranteeing delay jitter bounds in packet switching networks. In *Proceedings of Tricomm'91*, pages 35–46, Chapel Hill, North Carolina, April 1991.

113. C. A. Waldspurge. *Lottery and Stride Scheduling: Flexible Proportional -Share Resource Management*. PhD thesis, MIT, Laboratory of Computer Science, September 1995. MIT/LCS/TR-667.

114. C. A. Waldspurger and W. E. Weihl. Lottery scheduling: Flexible proportional-share resource management. In *Proceedings of OSDI 94*, pages 1–12, November 1994.

115. M. Waldvogel, G. Varghese, J. Turner, and B. Plattner. Scalable high speed routing. In *Proceedings of ACM SIGCOMM'97*, pages 25–36, Cannes, France, September 1997.

116. Z. Wang. User-share differentiation (USD) scalable bandwidth allocation for differentiated services, May 1998. Internet Draft, draft-wang-diff-serv-usd-00.txt.

117. W. E. Weihl. Transaction-Processing Techniques. *Distributed Systems, S. Mullender (ed.)*, pages 329–352, 1993.

118. W. Willinger, M. S. Taqqu, R. Sherman, and D. V. Wilson. Self-similarity through high-variability: Statistical analysis of ethernet lan traffic at the source level. In *Proceedings of ACM SIGCOMM'95*, pages 100–113, Boston, MA, August 1995.

119. D. Wrege, E. Knightly, H. Zhang, and J. Liebeherr. Deterministic delay bounds for vbr video packet-switching networks: Fundmental limits and practical trade-offs. *IEEE/ACM Transactions on Networking*, 4(3):352–362, June 1996.

120. D.E. Wrege and J. Liebeherr. A near-optimal packet scheduler for QoS networks. In *Proceedings of INFOCOM'97*, pages 576–583, Kobe, Japan, 1997.

121. J. Wroclawski. Specification of the controlled-load network element service, September 1997. Internet RFC 2211.

122. J. Yo. Scalable routing design principles, July 2000. Internet RFC 2791.

123. H. Zhang. *Service Disciplines for Integrated Services Packet-Switching Networks*. PhD thesis, University of California at Berkeley, Computer Science Division, November 1993. Technical Report UCB/CSD-94-788.

124. H. Zhang. Service Disciplines For Guaranteed Performance Service in Packet-Switching Networks. *Proceedings of the IEEE*, 83(10):1374–1399, October 1995.

125. H. Zhang and D. Ferrari. Rate-controlled static priority queueing. In *Proceedings of IEEE INFOCOM'93*, pages 227–236, San Francisco, California, April 1993.

126. H. Zhang and D. Ferrari. Rate-controlled service disciplines. *Journal of High Speed Networks*, 3(4):389–412, 1994.

127. L. Zhang. Virtual Clock: A new traffic control algorithm for packet switching networks. In *Proceedings of ACM SIGCOMM'90*, pages 19–29, Philadelphia, PA, September 1990.

128. L. Zhang, S. Deering, D. Estrin, S. Shenker, and D. Zappala. RSVP: A new resource reservation protocol. *IEEE Communications Magazine*, 31(9):8–18, September 1993.

129. Z.-L. Zhang, Z. Duan, L. Gao, and Y. T. Hou. Decoupling QoS control from core routers: A novel bandwidth broker architecture for scalable support of guaranteed services. In *Proceedings of ACM SIGCOMM'00*, pages 71–83, Stockholm, Sweden, September 2000.

130. Z.-L. Zhang, Z. Duan, L. Gao, and Y. T. Hou. Virtual time reference system: A unifying scheduling framework for scalable support of guaranteed services. *IEEE Journal on Selected Areas in Communication: Special Issue on Internet Quality of Services*, 18(12):2684–2695, Nov 2000.

ecture Notes in Computer Science

information about Vols. 1–2903

se contact your bookseller or Springer-Verlag